Southeast Asian Biodiversity in Crisis

The biodiversity of Southeast Asia is in great danger as a result of massive habitat modifications, forest fires and the overexploitation of wildlife. This book provides the first comprehensive review of the current state of Southeast Asia's terrestrial biotas and highlights the primary drivers responsible for the grave threat to the region's unique and rich biodiversity. The looming Southeast Asian biodiversity disaster demands actions. However, the region will continue to be constrained by socioeconomic variables (rampant poverty and lack of infrastructure) and any realistic solution must involve a multi-pronged strategy (political, socioeconomic and scientific) in which all major stakeholders (people, governments and national and international non-government organizations) partake.

NAVJOT S. SODHI is Associate Professor of Conservation Ecology at the National University of Singapore.

BARRY W. BROOK is Senior Research Fellow at the School of Environmental Research, Charles Darwin University.

CAMBRIDGE TROPICAL BIOLOGY SERIES

Southeast Asian Biodiversity

in Crisis

NAVJOT S. SODHI
National University of Singapore
BARRY W. BROOK
Charles Darwin University

CAMBRIDGE
UNIVERSITY PRESS

CAMBRIDGE UNIVERSITY PRESS
Cambridge, New York, Melbourne, Madrid, Cape Town,
Singapore, São Paulo, Delhi, Tokyo, Mexico City

Cambridge University Press
The Edinburgh Building, Cambridge CB2 8RU, UK

Published in the United States of America by Cambridge University Press, New York

www.cambridge.org
Information on this title: www.cambridge.org/9781107403130

First published 2006
First paperback edition 2011

A catalogue record for this publication is available from the British Library

ISBN 978-0-521-83930-3 Hardback
ISBN 978-1-107-40313-0 Paperback

CONTENTS

v

PREFACE

Worldwide, habitat loss, fragmentation and degradation are operating on a massive scale, and are accelerating. Many scientific studies have shown or predicted that this habitat destruction will have dire consequences for the future of global biodiversity. Habitat loss is particularly exceptional in Southeast Asia, which includes the nations of Brunei, Burma (Myanmar), Cambodia, East Timor, Indonesia, Laos, Malaysia, the Philippines, Singapore, Thailand and Vietnam (see Fig. 1.0). Deforestation rates in this biodiverse region are at least two times higher than other tropical areas. In light of the increasing threats to Southeast Asian biodiversity and the urgent need to make conservation decisions today, there is a critical need to contextualise the current state of Southeast Asian biodiversity, predict its future, determine possible ways to protect it and synthesise succinctly the current state of scientific research relevant to this region. These are the broad aims of this book. A brief organisational overview follows:

Chapter 1, Dwindling habitats. We report on deforestation rates of rain forest, and loss of other terrestrial habitats such as mangroves, within Southeast Asia. We compare the extent and rate of habitat loss in Southeast Asia with other tropical areas. We restrict our discussion and analyses to terrestrial habitats, as these have been more extensively damaged or lost during recent times.

Chapter 2, Biodiversity in a hotspot. We briefly discuss the biogeography of Southeast Asian biodiversity. We highlight the importance and uniqueness of Southeast Asian biodiversity, taking into consideration both animals and plants from a wide range of taxonomic groups. We show that due to high diversity and endemicity, Southeast Asian biodiversity is critical in the realm of global biodiversity.

Chapter 3, Biotic losses and other effects of habitat degradation. We report on the direct and indirect effects of habitat loss on the biodiversity in Southeast Asia, presenting a suite of relevant case studies. In addition, we

Figure 1.0 A map of Southeast Asia.

briefly review the characteristics that make species more or less extinction prone. We also explore the effects of habitat loss on the sanctity of ecological interactions and unique behaviours (e.g. mixed-species flocking).

Chapter 4, Beyond deforestation: additional threats to Southeast Asian biodiversity. In addition to habitat loss, Southeast Asian biodiversity faces substantial threats from anthropogenic overexploitation, such as harvesting for wild bushmeat and removal for the pet trade. We highlight the effects of human exploitation on biodiversity in the region. For example, we illustrate the unsustainable harvesting of species such as the Maleo (*Macrocephalon maleo*) and edible-nest swiftlets (*Collocalia* spp.). One of the other consequences of habitat degradation and human settlement is the spread of invasive species. We discuss the documented and potential effects of invasive species on the native biodiversity of Southeast Asia.

Chapter 5, The projected future of biodiversity in Southeast Asia. Based on the best available data, published literature and our own quantitative analyses, we present model scenarios for the future of biodiversity in the region. These projections are based on reported biotic extinctions, including losses of local populations, and biogeographically defined rates of deforestation. Detailed projections are made for selected taxonomic groups (e.g. birds) for which sufficient data are available. However, such projections give indications of the future of Southeast Asian biodiversity in general under a range of plausible prospective scenarios.

Chapter 6, Challenges and options for conservation in the region. Here we attempt to provide realistic recommendations for the protection of existing biodiversity. We stress the need to integrate the major social issues (e.g. human hunger) in order to achieve tangible conservation.

Our emphasis is on the biological, rather than social or political components of the conservation debate. The content should appeal to advanced undergraduate and graduate students, scientists, and managers with an interest in conservation biology of the Southeast Asian region specifically, and tropical areas in general. Because of a paucity of Southeast Asian examples and context in other biodiversity reference books, such as *Biodiversity II: Understanding and Protecting our Biological Resources* (Reaka-Kudla *et al.* 1997) or *The Biodiversity Crisis: Losing What Counts* (Novacek 2001) and text books such as *A Primer of Conservation Biology* (Primack 2004) or *Fundamentals of Conservation Biology* (Hunter 2001); the present title will complement both these types of books. We hope you find *Southeast Asian Biodiversity in Crisis* both useful and interesting.

ACKNOWLEDGEMENTS

We are indebted to Lian Pin Koh, Neil F. Ramsay and Malcolm Soh for their invaluable assistance. Thanks go to Tom Brooks, Richard Corlett, Rob Dunn, Bruce Campbell, Ruth O'Riordan, Benito Tan, Neil F. Ramsay, Penny Van Oosterzee and Lian Pin Koh for reading drafts. Lian Pin Koh, Tien Ming Lee, Tommy Tan and Arvin Diesmos provided some of the photographs. We both thank our families for support.

Chapter 1

Dwindling habitats

If the human impact on the natural environment continues unabated at its present rate, or indeed increases in severity, then by the turn of the next century, the resulting changes in land use will have exerted a profound and irreversible effect on tropical biodiversity (Sala *et al.* 2000). Habitat loss will probably have far greater effects on terrestrial ecosystems in the tropics than other drivers such as climate change, elevated carbon dioxide (CO_2) levels and invasive species (Sala *et al.* 2000). Rain forest loss, its degradation and fragmentation, has been a widely publicised example of habitat loss in the tropics. Anthropogenic activities such as logging are degrading and destroying tropical rain forests at a rate that lacks historical precedence (Jang *et al.* 1996; Whitmore 1997; Laurance 1999). In the decade from 1979 to 1989, the annual global area of tropical forest lost increased by more than 90% (Myers 1991). Given that the vast majority of the Earth's terrestrial biodiversity is harboured in these threatened and little studied biomes (Wilson 1988; Myers *et al.* 2000; Sodhi & Liow 2000), they represent obvious foci for conservation. In this chapter, we discuss the loss of native habitats (primary forests) of Southeast Asia.

Unprecedented losses

Almost the whole of Southeast Asia was covered by forest 8 000 years ago (Billington *et al.* 1996). Today, this region has the highest rate of rain forest loss, with deforestation rates more than double those of other tropical areas (Hannah *et al.* 1995; Laurance 1999; Achard *et al.* 2002), and only a few areas (e.g. Borneo and Sulawesi) retain large tracts of intact primary forests (Laurance 1999). Using the United Nations Food and Agriculture Organization's (FAO) data on forest cover change from 1980 to 1990 (FAO 1993), Laurance (1999) estimated that 15.4 million ha of tropical forest is destroyed every year, with an additional 5.6 million ha being degraded through such activities as selective logging. Overall, an average of 1.2% of existing tropical forests are degraded

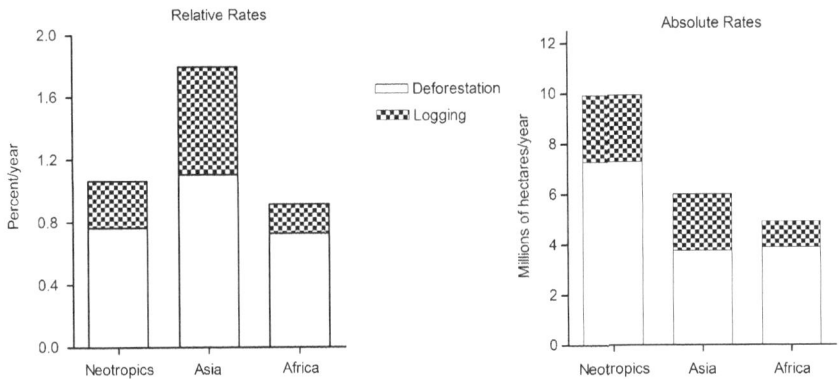

Figure 1.1 Relative and absolute rates of forest conversion in the
major tropical regions throughout the 1980s. (Modified from Laurance
1999. With permission from Elsevier.)

or destroyed every year (Whitmore 1997; Laurance 1999). In terms of absolute
loss of area, forest conversion is the highest in Neotropics (10 million ha/year)
followed by Asia (6 million ha/year) and Africa (5 million ha/year). However, if
we consider forest conversion relative to the existing forest cover in the region,
Asia clearly tops the list (Laurance 1999) (Fig. 1.1), with 1.5 million ha of forest
removed each year from the four main Indonesian islands of Sumatra,
Kalimantan (Indonesian Borneo), Sulawesi and Irian Jaya (Indonesian New
Guinea) alone (DeFries *et al.* 2002). Even the so-called 'protected forests' of
Kalimantan declined by more than 56% (or 2.9 million ha) between 1985 and
2001 (Curran *et al.* 2004).

There is a controversy as to whether the FAO values are accurate, as they
may fail to include catastrophic events such as the vast 1997–8 forest fires in
Indonesia, and perhaps erroneously include plantations as forest cover
(Matthews 2001; Achard *et al.* 2002). Deploying remotely sensed satellite
imagery, Achard *et al.* (2002) reported that tropical forest loss may be much
lower (5.8 million ha/year) than FAO estimates. Yet even Achard *et al.*'s
estimates have been questioned. It has been argued that their lower estimates of forest loss may be due to lack of representativeness owing to their
relatively small sample sizes (Fearnside & Laurance 2003). Nevertheless,
despite the different methodology used, Achard *et al.* (2002) also found, as
reported earlier by Laurance (1999), that rates of deforestation and forest
degradation are among the highest in Southeast Asia (Fig. 1.2).

The main culprit in this devastating forest loss in Southeast Asia is expansion
for agriculture, with more than 1 million ha of forest converted annually by this
human activity (Achard *et al.* 2002). A particularly worrying trend is that

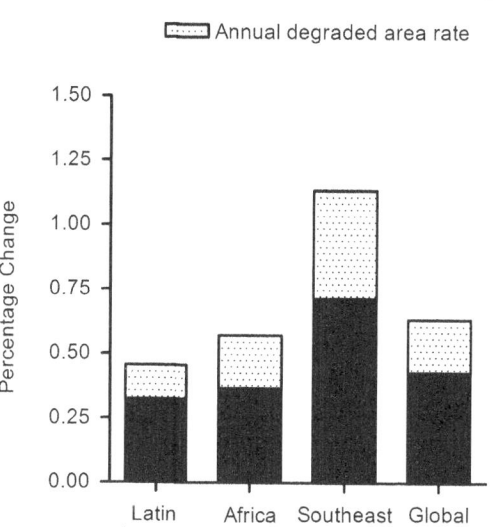

Figure 1.2 Mean annual estimates of deforestation in the humid tropics during 1990–7. (Data from Achard *et al.* 2002.)

although native forest loss seems to be decelerating with time in tropical Latin America, it continues to accelerate in tropical Asia (Matthews 2001) (Fig. 1.3).

Globally, 0.8% of native tropical forests (primary and secondary forest, excluding plantations) are likely to be lost each year (Matthews 2001). Let us now consider what is happening in different Southeast Asian countries. In 1880, almost all Southeast Asian countries had more than 70% of the original forest cover intact (Flint 1994) (Fig. 1.4).

The island nation of Singapore was an exception, with only 30% of its forest intact at that time, because as early as the late nineteenth century, it supported a relatively high population density (2 persons/ha) and was a well developed international trading centre of the British Empire (Saw 1970). However, even Singapore was almost totally covered by rain forest in 1819, yet now less than 1% remains (Turner *et al.* 1994). Although Singapore represents an apex of forest loss and urbanisation in the region, other countries are moving in the same direction. Between 1961 and 1991, forest cover declined from 53% to 27% in Thailand (Ruangpanit 1995) (Fig. 1.5), and in the Philippines, forest cover has been reduced by 55% since 1948 (Kummer & Turner 1994).

Perhaps most dramatically, it has been estimated that by 2010, human actions will cause the near complete destruction of native lowland forests

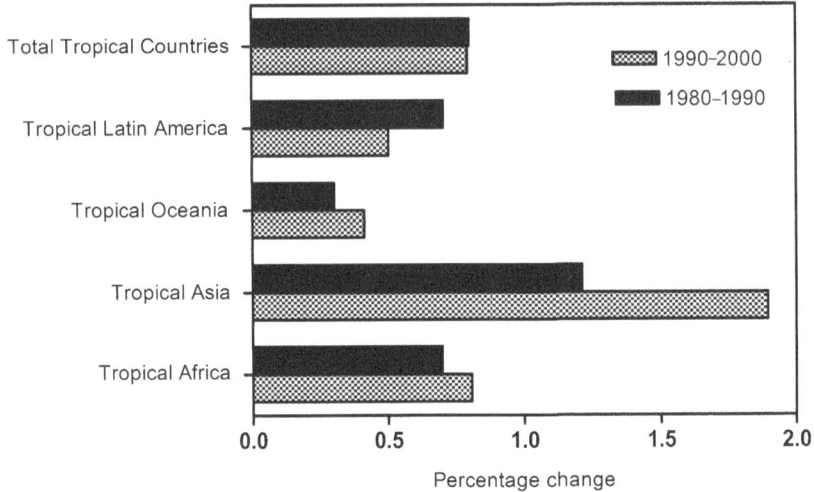

Figure 1.3 Worsening deforestation rates in all tropical regions except Latin America. (Data from Matthews 2001.)

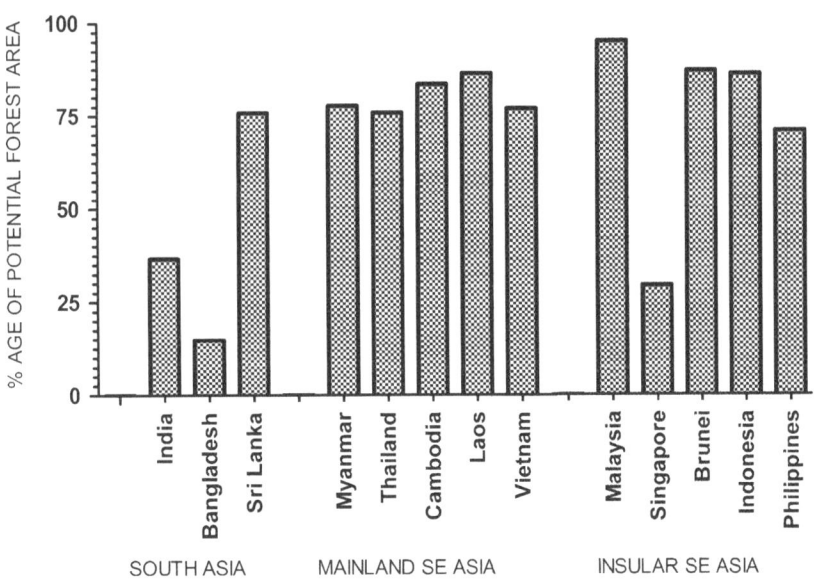

Figure 1.4 Percentage of potential forest area within each South and Southeast Asian nation that was still occupied in 1880 by forest. (Modified from Flint 1994. With permission from Elsevier.)

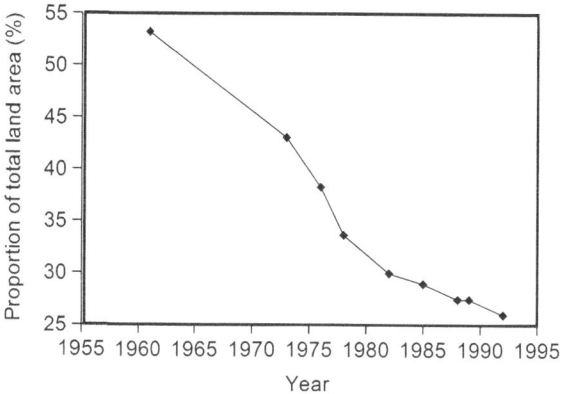

Figure 1.5 Periodic change in forest area in Thailand from 1961 to 1991. (Data from Ruangpanit 1995.)

(<1000 m elevation) from the hyper-biodiverse regions of Sumatra and Kalimantan (Jepson *et al.* 2001). Such a massive loss of habitat will almost certainly have profound knock-on effects for the region's spectacular mega-fauna, such as the Sumatran rhinoceros (*Dicerorhinus sumatrensis*), Sumatran tiger (*Panthera tigris sumatrae*) and Asian elephant (*Elephas maximus*).

Vietnam is the only country in the region with a net annual increase in forest cover. Establishment of plantations helps offset annual losses of natural forest cover in Vietnam of the order of 30 000 per ha year[-1] (http://www.fao.org/DOCREP/004/Y1997E/y1997e0t.htm#bm29).

According to the World Resource Institute (WRI) estimates (WRI 2003), annual deforestation rates vary between 0 (Singapore) and 2.9% (Thailand) (see Table 1.1), and the various countries in Southeast Asia currently retain between 0.3% (Singapore) and 65.5% (Cambodia) of their original forests. Although the lowest estimates of remaining forest cover were identical for both WRI and FAO, the latter organisation reported the highest forest cover to be remaining in the tiny nation of Brunei (83.3%; Table 1.1), located in northern Borneo. Differences in definitions and methods appear to have led to the differences between the two data sets.

The lowland rain forests of Southeast Asia are particularly imperiled due to their high accessibility to humans, and are increasingly being converted to logging concessions, agricultural land and urban areas (Kummer & Turner 1994). In addition to this wide-spread forest type, other rain forest types are by no means being spared. For example, montane/submontane rain (cloud) forests (usually >1000 m elevation) provide timber, fuel-wood, soil and watershed protection. There is about 50 million ha of montane forest worldwide, and it is currently being cleared at an average rate of twice the global deforestation rate

Table 1.1. *Summary information on forestry and biodiversity in Southeast Asia, showing land area, original and current forest area, and change in forest area*

Country	Land area (ha)	Original forest area (ha) (% land area)	Current natural forest area (000 ha) (% original forest area)		Mean annual % change in forest area 1990–2000	
			WCMC	FAO	'Natural forest' – WRI	'Total forest' – FAO
Myanmar	65755	65755 (100)	33519 (51.0)	33598 (51.1)	− 1.5	− 1.4
Laos	23080	23057 (99.9)	4495 (19.5)	12507 (54.2)	− 0.5	− 0.4
Vietnam	32550	32452 (99.7)	5015 (15.5)	8108 (25.0)	− 0.3	0.5
Thailand	51089	51089 (100)	17107 (33.5)	9842 (19.3)	− 2.9	− 0.7
Cambodia	17652	17652 (100)	11562 (65.5)	9245 (52.4)	− 0.6	− 0.6
Malaysia	32855	32691 (99.5)	13452 (41.1)	17543 (53.7)	− 1.4	− 1.2
Singapore	54	54 (100)	0.2 (0.3)	0.2 (0.3)	0	0
Indonesia	181157	181157 (100)	91134 (50.3)	95116 (52.5)	− 1.5	− 1.2
Brunei	527	527 (100)	267 (50.7)	439 (83.3)	− 0.3	− 0.2
Philippines	29817	28416 (95.3)	2405 (8.5)	5036 (17.7)	− 2.1	− 1.4
Total	434536	432850 (99.6)	178956 (41.3)	191434.2 (44.2)	− 1.4	− 1.0

Data taken from FAO, United Nations Environment Programme, World Conservation Monitoring Centre (WCMC) and World Resources Institute (WRI). (Reprinted from Sodhi *et al.* 2004b. With permission from Elsevier.)

(Long 1994; IUCN 2000). Because of their unique environmental conditions (e.g. low humidity, cooler temperatures), montane forests support a high degree of endemism. For example, the proportion of endemic moths is reported to be at least twice as high in montane forests than in their lowland counterparts in Sarawak (Lanjak-Entimau) (Chey 2000). Montane forests also have a low recovery potential following disturbance (Ohsawa 1995). Yet despite their fragility and high endemism, human activities continue to threaten these vulnerable forests (Ohsawa 1995; IUCN 2000).

Some 85% of global forest loss occurs in the tropical rain forests (Whitmore 1997). However, Southeast Asia also contains seasonal deciduous forests. These generally lie below 1000 m elevation in certain countries such as Thailand (Ruangpanit 1995), and constitute only 7% of the existing forests in Asia. Due to their close proximity to humans, seasonal forests also suffer from a similar predicament as lowland rain forests in Southeast Asia. In fact, seasonal forests are often pooled for convenience with rain forests, and are thus included in some of the regional deforestation calculations (Achard *et al.* 2002). It is estimated that the seasonal forests, on their own, are being lost at the rate of 1.4% per year (Whitmore 1997).

Table 1.2. *Estimates of mangrove area and area loss for selected countries in Southeast Asia*

Country	Mangrove area (km^2)	Approximate % lost	Period covered
Brunei	171	29	Original extent to 1986
Cambodia	851	—	—
Indonesia	42 550	55	Original extent to 1980s
Malaysia	6424	74	Original extent to 1992–3
Myanmar	3786	75	Original extent to 1992–3
Philippines	1607	67	1918 to 1987–88
Singapore	6	—	—
Thailand	2641	84	Original extent to 1993
Vietnam	2525	37	Original extent to 1993

Adeel & Pomeroy (2002). Copyright Springer-Verlag.

Mangrove forests represent another unique tropical ecosystem. Mangroves are juxtaposed between land and sea and are found within 25° north and south of the equator, and Southeast Asia supports 40% of the world's total mangrove cover (Sasekumar *et al.* 1994). In addition to over-harvesting, mangroves face threats from other factors such as pollution, silting, coastal development, aquaculture development, and boating and shipping (Adeel & Pomeroy 2002). Traditionally, mangroves have been undervalued and largely considered to be useless swamps or wasteland (Liow 2000; Adeel & Pomeroy 2002). However, as with other forest types, mangroves have both biodiversity and utilitarian values. The presence of mangroves may enhance fish, shrimp and prawn catch (Baran & Hambrey 1998). It is estimated that fisheries related annual income from 1 ha of mangrove can range from US$66 to almost US$3000 (Baran & Hambrey 1998). Although this estimate may be inflated as it does not include fisheries yield exclusively reliant on mangroves, it does show that livelihoods do depend on this habitat type. Because of their other benefits (e.g. ecosystem services needed for the maintenance of offshore fisheries), conversion of mangroves for aquaculture actually generate around 70% less revenue from the overall system than if they had been left in a pristine state (Balmford *et al.* 2002). Despite their environmental and economic benefits, mangroves are currently being lost at a rate of 2–8% per year, with between 29 and 84% loss of the original mangrove cover in Southeast Asian countries (Adeel & Pomeroy 2002) (Table 1.2).

Singapore epitomises mangrove destruction and conversion in Southeast Asia, supporting 6334 ha (63% of original mangrove forest cover) in 1953, but only 6.5% by 1993 (Hilton & Manning 1995), and a further projected reduction to 4% by 2030 (Fig. 1.6). The primary driver of this massive loss

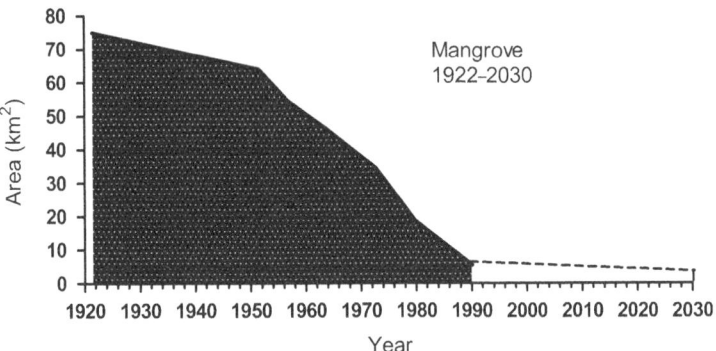

Figure 1.6 Change in area of mangrove forest in Singapore, with estimates for the year 2030. (Reprinted from Hilton & Manning 1995. With permission from Cambridge University Press.)

of mangrove forests in Singapore has been coastal developments associated with urban expansion and industrialisation (Hilton & Manning 1995). The mangrove loss in Singapore has certainly resulted in biotic losses. For example, at least four mangrove plant species (e.g. *Barringtonia conoidea*) have been extirpated from the island (Liow 2000).

The pathology of habitat loss

Direct causes of deforestation (and loss of other habitats) are many, including slash and burn clearing, selective logging, cattle ranching, plantations, agriculture, fuel-wood collection and transmigration. These drivers can act singly or in concert. In Asia, the main proximate drivers of deforestation are agriculture, followed by wood extraction and infrastructure expansion (Geist & Lambin 2002). The precise underpinnings of these causes of deforestation are complex, however. For instance, some governments have little choice but to sell forests as logging concessions to alleviate foreign debt (Bawa & Dayanandan 1997). Below we discuss in some detail the various drivers of habitat loss in the Southeast Asian region.

Human population pressure

Resource consumption by humans shows no sign of abating. Rapid economic development, population expansion and poverty are key drivers of land conversion (Giri *et al.* 2003). In the century from 1880 to 1980, the human population of Southeast Asia has increased from 0.1 to 0.8 person/ha

(Flint 1994). Within the next 100 years, it is likely that as many as 11 billion people will inhabit the planet, a number that will be difficult to sustain (Palmer *et al.* 2004). Urbanisation will greatly expand in the future, with expectations that more than half of the world's total human population will be living in cities by 2030 (Palmer *et al.* 2004). Expanding human population and specifically its actions (e.g. land conversion), exerts pressure on the native biodiversity (Cardillo *et al.* 2004).

It would be immeasurably informative, from both a scientific and management perspective, if we could hypothetically excise a representative Southeast Asian country, allow it to fulfil its economic potential, and document the consequent losses of natural habitats and biodiversity, all within a greatly accelerated time frame. It is perhaps both depressing and fortunate that Singapore is exactly such an ecological worst case scenario for Southeast Asia. Singapore has experienced an exponential population growth from around 150 subsistence-economy villagers around 1819 to 4 million people in 2001 (Corlett 1992; WorldBank 2003). In particular, Singapore has transformed itself from a third world country of squatters and slums to a first world metropolis of economic prosperity within the past few decades, and has thus been widely regarded by the regional developing countries as the ideal economic model. However, the success of Singapore came with a hefty price, one that was unfortunately paid for most heavily by its biodiversity (see Chapter 3 for a detailed analysis). The island has suffered massive deforestation, initially from the cultivation of short-term cash crops (e.g. gambier: *Uncaria gambir*, rubber: *Hevea brasiliensis*), and subsequently from urbanisation and industrialisation (Corlett 1992). Similar environmental scenarios are already unfolding in other Southeast Asian countries (Jepson *et al.* 2001). As the human population of Southeast Asia continues to grow, enormous pressures will be placed upon its natural resources (WorldBank 2003). The current trend in Southeast Asia suggests that forest loss is likely to increase in step with both human population density and economic expansion (Fig. 1.7). The point to note, however, is that population pressure represents only one of the factors in habitat loss, even with the relatively low populated areas of Southeast Asia, there is a widespread loss of natural forests (Whitmore 1997).

Burgeoning human population means more mouths to feed. Agriculture is the main factor in land conversion in the tropics, with an estimated contribution to annual tropical forest losses of as high as 90% (Hardter *et al.* 1997; Achard *et al.* 2002). In Asia, 100 million ha of land was converted for cropland between 1880 and 1980. Over these 100 years, the area of land converted for agriculture increased by four-fold in Southeast Asia (Flint

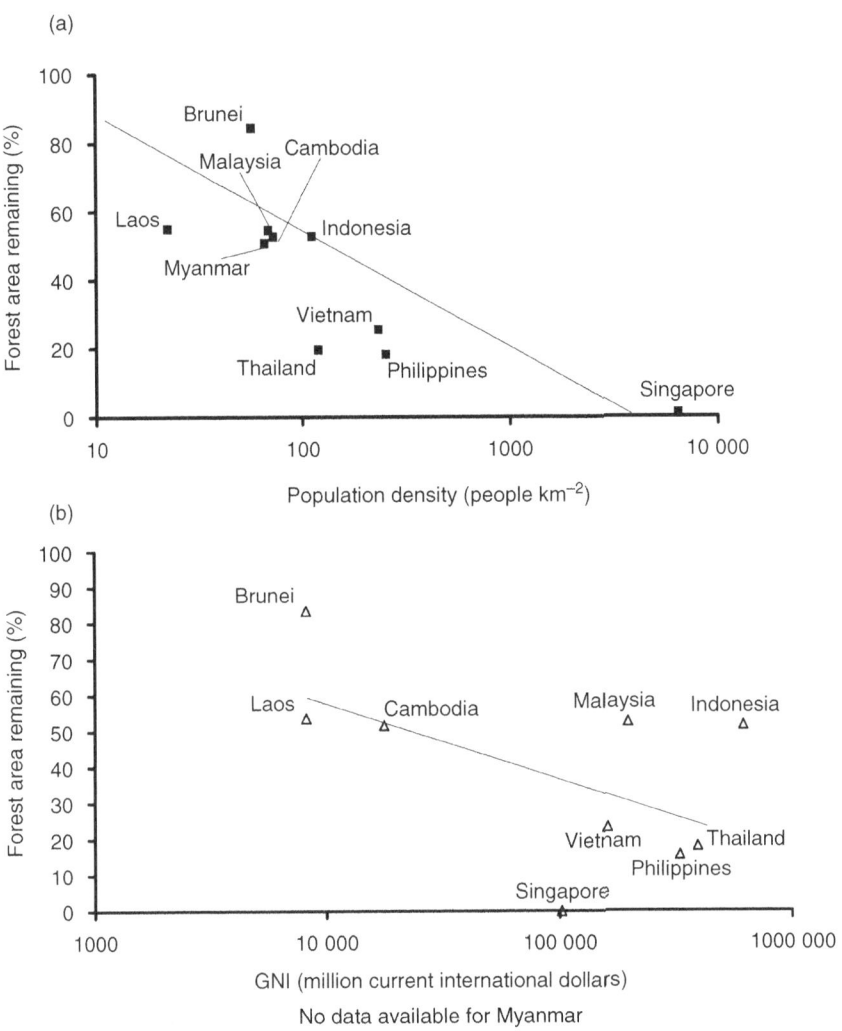

Figure 1.7 Socioeconomic correlates of forest loss. The proportion of forest area remaining in Southeast Asian countries correlated with (a) Population density ($r = -0.78$, $p = 0.008$) and (b) Gross National Income (GNI) at current international dollars ($r = -0.57$, $p = 0.111$) in 2000. GNI of Brunei was taken in 1998. (Reprinted from Sodhi et al. 2004b. With permission from Elsevier.)

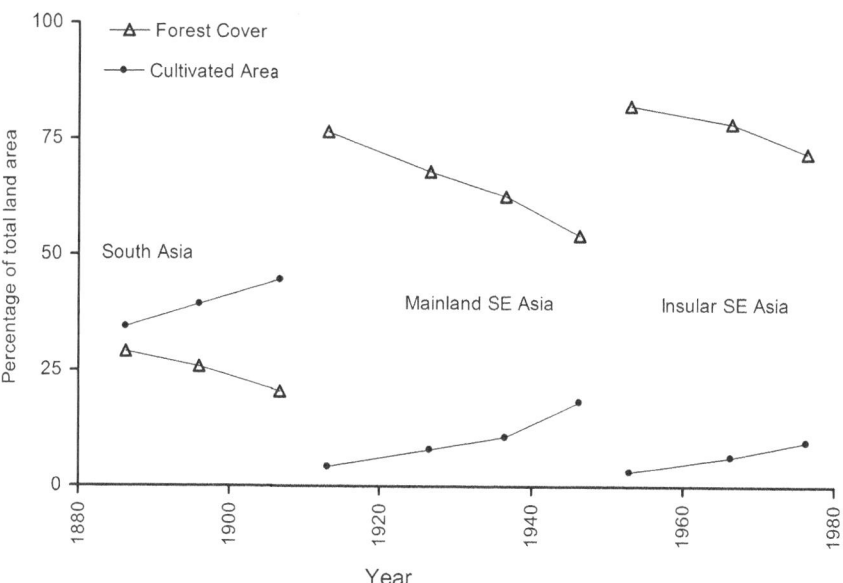

Figure 1.8 Change in cultivated area and forest cover of South and Southeast Asian countries (modified from Flint 1994). With permission from Elsevier.

1994; Richards & Flint 1994), largely at the expense of its forests (Flint 1994; Bawa & Dayanandan 1997) (Fig. 1.8).

By 2030, it is predicted that an additional 120 million ha of agricultural land will be needed by developing countries to support their increased populations (Jenkins 2003). Therefore land clearing for agriculture is almost certain to continue at a pace. Further, land conditions in many areas of Southeast Asia are not particularly conducive for sustainable agriculture, due to factors such as low soil fertility and high levels of erosion, thus promoting a cycle of forest destruction. Farmers have to burn the forest vegetation in order to release nutrients to enhance soil fertility. These nutrients are usually washed away quite rapidly due to the region's characteristically high rainfall, which in turn makes the soil less fertile: in as little as three years the ground is no longer capable of supporting crops (Hardter *et al.* 1997). Farmers are then forced to look for new forested areas to clear and burn because of deteriorated soil fertility.

Perverse subsidies

Some governmental actions promote poor land use practices. Subsidies designed to promote agricultural production also act to facilitate land clearing (Barbier 1993; James *et al.* 1999). For example, during the 1990s, Asian governments, with the support of the International Development Bank, promoted intensive coffee (*Coffea robusta*) cultivation in countries such as Indonesia and Vietnam, elevating Indonesia to the world's fourth largest coffee exporter and the second largest producer of *C. robusta* after Vietnam. This move resulted in massive forest conversion, but eventually proved to be economically unsustainable due to overproduction and subsequent price collapse (O'Brien & Kinnaird 2003). Despite it being clear that coffee production is not economically attractive and that it is a detriment to biodiversity, there are plans to further expand coffee production by the Indonesian government. O'Brien & Kinnaird (2003) recommended that coffee cultivation should be strongly discouraged in protected areas, and strident attempts should be made to curtail deforestation due to this type of cultivation. Certainly, organizations such as the International Coffee Organization need to play a bigger role in promoting a balance between coffee production and biodiversity needs.

In addition to agriculture, government finance of road construction can act as an indirect subsidy to facilitate logging (Flint 1994). Such road construction can also elevate the hunting of wild animals by providing easier access to the forests (see Chapter 4). In addition to road construction, under-pricing of timber and subsidising (e.g. low logging fees and taxes) of private harvesting assist in deforestation (Barbier 1993). For example, in the Philippines, timber revenues collected by the government were six times lower than those that should have been collected according to prevailing market value (US$39 versus US$250 million; Barbier 1993).

Massive resettlement programmes, as have been carried out in Indonesia, Thailand and the Philippines also facilitate deforestation (Bryant *et al.* 1993). Last but not least, the liberal granting of forestry concessions, largely though cronyism and corruption, does not bode well for the remaining tropical forests (Geist & Lambin 2002; and see Chapter 6).

Commercial logging

Commercial logging is another common driver of deforestation. Logging activities account substantially in tropical deforestation (Geist & Lambin 2002). Trees are cut for sale as timber, timber products (e.g. woodchips) or pulp. Forestry industries account for 3% of the world's gross economic output or approximately US$330 billion annually (Sizer & Plouvier 2000). The total annual worldwide consumption of wood for production is around 1.5 billion cubic metres, with a comparable amount consumed as firewood. The Asia–Pacific region champions log exports in the tropics, encompassing 67% of the total volume (Sizer & Plouvier 2000) (Fig. 1.9).

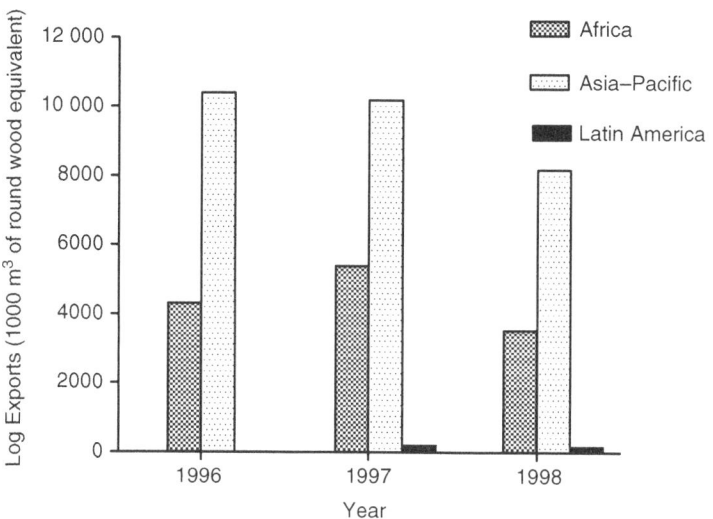

Figure 1.9 Comparison of log exports among tropical regions of the world.
(Data from Sizer & Plouvier 2000.)

Since the 1950s, Southeast Asia remains the major exporter of wood
and wood products (Ooi 1990). Over the past couple of decades, Malaysia
and Indonesia have produced more than half of the world's commercial tropical timber, with Thailand also in the top five nations (ITTO 2003) (Fig. 1.10).

Countries such as Japan, South Korea and People's Republic of China
are the main importers of tropical timber products (Sizer & Plouvier 2000).
Forest products exported by the developing countries are usually subjected
to low export duties, particularly on unprocessed logs (Barbier 1993).
Trade-liberation through the removal of export restrictions, may increase
log exports (by up to four-fold in the Philippines) and thus cause a further
acceleration in deforestation (Barbier 1993). In addition, the current bans on
commercial logging in India and the People's Republic of China places an
added demand for the supply of Southeast Asian wood (http://www.birdlife.
net/action/science/species/asia_strategy/pdf_downloads/forestsFO4.pdf).

Deforestation thus is promoted by governments in Southeast Asia due
to high international demand for tropical wood and wood products, with
up to 83% of global tropical wood products being sourced from this region
(Kummer & Turner 1994). Examples of sustainable natural forest utilisation in the tropics are difficult to find (Bowles et al. 1998; Putz et al. 2000;
Laurance et al. 2001). One of the reasons for this is that unsustainable
logging remains 20–450% more profitable, at least over the short-term,
than the sustainable practices (Bowles et al. 1998; see Balmford et al.

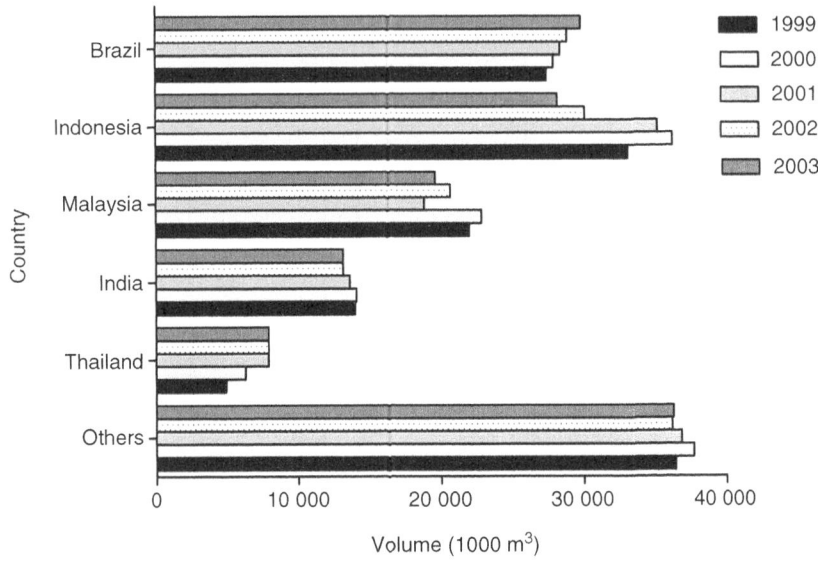

Figure 1.10 Major tropical timber producers. (Source ITTO 2003.)

2002). Most of the logging in Southeast Asia is still done by clear-cutting (Barbier 1993), yet even 'selective logging' can be very wasteful, resulting in the felling of an average of 25 non-commercial trees for every commercial-quality tree extracted (Myers 1991), and 40–50% of the canopy cover being destroyed (Cochrane 2003). Further, logging concessions are typically awarded for short durations (e.g. 5–25 years), which tends to promote logging of new areas (Barbier 1993). In the Malay Archipelago, 85% of the logging occurred in primary forests from 1981 to 1990 (Whitmore 1997). Throughout Southeast Asia, old-growth forests are being depleted rapidly (Barbier 1993). Clearly, stiff guidelines are needed to regulate and perhaps curtail mass tropical timber production and export. These might include mandatory timber certifications, economic incentives to better encourage sustainable harvesting, substantial reviews of export duties and laws, and a more wide-spread promotion and education on environmental issues (see Chapter 6). Further, because they are often the only institutions in remote areas, logging companies can themselves assist in environmental protection, though officials wanting to stop illegal logging are intimidated with arson and murder in certain instances (Jepson *et al.* 2001).

Agriculture and commercial logging may have differing impacts on deforestation rates in Southeast Asia. However, in many areas, both logging and agriculture act in concert to exacerbate deforestation (Kummer &

Turner 1994). Logging operations, as mentioned earlier, enhance physical access to forests by, for example, creating roads. This increased access increases the chances of invasion by both humans (e.g. hunters, farmers and miners) and exotic organisms associated with humans, and is a cause of considerable concern for the long-term prospects of the residual biodiversity (Laurance et al. 2001).

Poor institutions and political will

Anaemic national institutions and poor enforcement of legislation remain a major hindrance to abating tropical deforestation (Laurance 1999). Liberal granting of forest concessions, non-existent or poor forestry practices, weak governance structures and political corruption, all work to exacerbate deforestation in developing countries (Bryant et al. 1993; Geist & Lambin 2002; Smith et al. 2003). Illegal logging and encroachment into nature reserves remains a problem across Southeast Asia (http://www.fao.org/forestry; Whitten et al. 2001). For example, it is reported that illegal and possible unsustainable logging remains rampant in Indonesia, with the implicit backing of politicians, businesses and the military (Kinnaird & O'Brien 2001; Whitten et al. 2001; Stibig & Malingreau 2003; see Chapter 6). Certain people implicated in illegal logging continue to retain prominent political positions (Whitten et al. 2001). Some government officials that attempt to stop illegal logging in Indonesia are seriously intimidated with arson, attempts at bribery and murder (Jepson et al. 2001). In addition, forest policies are not sufficiently well developed to adequately protect the remaining forest. Indonesia's paper and pulp industry has grown by seven-fold since the 1980s, with similar expansion in the production of plywood. Further, the oil palm (Elaeis guineenis) plantation areas and resettlement plans for some native people have placed a further pressure on the forests (Stibig & Malingreau 2003).

What is happening in Indonesia is not an anomaly. It epitomises disturbing actions across Southeast Asia. Illegal logging is rampant in the region (Barbier 1993), with the volume of illegal logging output rivaling that of legal logging. Further, the construction of roads facilitates illegal logging in already over-logged areas (Barbier 1993). In many Southeast Asian nations, forests are managed by government departments whose main emphasis seems to be to increase commercial logging and export of timber products (Byron & Waugh 1988). Further, these departments remain understaffed and politically influenced (Barbier 1993). Logging companies identified as having poor environmental practices (e.g. promoting forest fires) continue to carry out business as usual (Whitten et al. 2001). Rampant corruption and illegal activities hinder proper management of forests throughout Southeast Asia (Kummer & Turner 1994; see Chapter 6). Corruption has been a major contributing factor in the heavy deforestation of the Philippines (Kummer & Turner 1994), and this is very likely to also be a major factor promoting deforestation in other Southeast Asian countries. The few well publicised struggles by some native groups (e.g. Penan of Sarawak) to halt logging have generally been in vain (Bryant *et al.* 1993).

Invaluable losses

Much has been written on the biodiversity, economic and ecosystem aspects of tropical habitat loss, and in particular rain forests. Below we summarise and largely reiterate what has been said to date, but attempt to provide a more explicit Southeast Asian context, where possible.

Environmental filters

Our atmosphere works like a greenhouse. Short-wave radiation emitted by the sun passes through the atmosphere to earth, and some of the sun's energy is radiated back into the atmosphere in the form of long-wave infrared radiation. Certain trace gases such as CO_2, methane and nitrous oxide (N_2O) trap some portion of this infrared radiation (Houghton 2004).

Over most of the past few hundred million years, forests of some description have helped to maintain the balance of carbon in the atmosphere. Forests convert carbon into cellulose and release oxygen through photosynthesis. Through photosynthesis, forests act as carbon sinks as they absorb and store atmospheric carbon (Page *et al.* 2002). However, as detailed in the earlier sections of this chapter, a sizeable fraction of the world's tropical forests have been, and continue to be, depleted by recent human activities. This trend, coupled with the release of greenhouse gases through the burning of fossil fuels for industry, is jeopardising the atmospheric balance. After fossil fuel consumption, human modification of vegetation and soils are the next main causes of global carbon emissions (Flint 1994). Increases in atmospheric CO_2 is estimated to be alone responsible for about half of all global warming. The plants and soil harbour between 460 and 575 billion metric tonnes of carbon (Flint 1994). Tropical deforestation contributes significantly towards total CO_2 emissions (Flint 1994).

When a forest is felled and burnt for the establishment of cropland and pastures, the stored carbon is combined chemically with oxygen and released into the atmosphere. Releasing CO_2 in the atmosphere enhances the greenhouse effect, and almost certainly contributes to elevating global temperatures (IPCC 2001). Tropical forests particularly have high potential for carbon storage (Malhi & Grace 2000). Mean carbon fluxes have increased from 0.6 petagrams (Pg)/year to 0.9 Pg/year from the 1980s to 1990s because of tropical deforestation (taking forest regrowth into account) (DeFries et al. 2002), although the precise figures remain the subject of debate. In Southeast Asia, net carbon flux for 1 ha of forest is 151 tonnes of carbon (Achard et al. 2002). It is estimated that deforestation in Southeast Asia releases 465 million tonnes/year of carbon emissions to the atmosphere (Phat et al. 2004). This represents 29% of global carbon release due to deforestation. Globally, temperatures will be 1–2 °C warmer in 2050 than today (Jenkins 2003). Total CO_2 emissions might increase by 100–200 times by that date, largely because of the continued burning of fossil fuels, clearance of native vegetation and release of the carbon it had previously sequestered (Jenkins 2003). Deforestation thus facilitates global warming by eliminating a major potential sink of atmospheric CO_2 and releasing its stored carbon into the atmosphere.

Tropical peatlands harbour large quantities of terrestrial carbon (Page & Rieley 1998). Natural peatlands can have up to 20 m thick peat deposits with swamp forests. Drainage and forest clearing increases substantially the susceptibility of peatlands to fire. During the forest fires in Indonesia during 1997–8, an estimated 1.45 million ha of peatlands were burnt, comprising 300 000 ha in Sumatra, 750 000 ha in Kalimantan and 400 000 ha in Irian Jaya (Page & Rieley 1998). It is estimated that during this event 0.48–0.56 gigatonnes (Gt) of carbon was released through peat burning, with additional 0.05 Gt released from burning of overlaying vegetation (Page & Rieley 1998). This release represents 13–40% of the mean annual global carbon emissions from fossil fuels, and is the largest single annual increase since records began in 1957. The earlier 1982–3 fires that spread across the East Kalimantan forests of Borneo may have been even more dramatic, burning an estimated 2–3 million ha of tropical forest (Hartshorn & Bynum 1999).

Methane is another greenhouse gas that is generated by wetlands, paddyfields, herbivorous animals (e.g. termites and ruminants) and through the combustion of fossil fuels. Forests in some areas (e.g. Java) are cleared extensively for rice (*Oryza sativa*) cultivation. Since large land areas of Southeast Asian countries are now under rice cultivation (44 million ha) (FAO 2004), the amount of methane release represents a real concern. World methane emissions have increased by 49% from 1940 to 1980, and it is estimated the paddy cultivation contributed 83% to this increase (Bolle *et al.* 1986). Although the exact amount of methane emission through rice cultivation remains debatable, it is clear that expansion in paddy cultivation is a major potential source of elevated methane emissions. Forest canopies also absorb another greenhouse gas, N_2O, and other atmospheric pollutants, and thus help maintain air quality (Chivian 2002).

Precipitation and temperature regulation

Tropical forests are vital mediators of both local and regional climates. They act as heat pumps by distributing solar radiation from the equator to temperate zones, thus warming the temperate areas but cooling the tropics. Hence, deforestation can elevate air temperatures in the tropics (Berbet & Costa 2003). Tropical forests also affect local and regional precipitation levels by releasing large amounts of water through evaporation and evapo-transpiration. This water generates clouds and precipitation. It is predicted that deforestation may result in an 8% decline in precipitation in Southeast Asia over the next decade, with a much steeper precipitation decline of 17% in Indonesia alone as it contains substantial forest cover (Hoffman *et al.* 2003). Another study predicted that annual precipitation will be reduced by as much as 172 mm in Southeast Asia due to deforestation (Zhang *et al.* 2001). Thus, this reduction in precipitation will enhance drought conditions, and make forests and peatlands ever more susceptible to fires, creating a positive feedback loop. Fires in turn can affect biodiversity (see Chapter 4).

Protected watersheds and soils

Forests can assist in regulating the water flow to downstream areas. Thus deforestation can alter the natural water flow resulting in either flood or drought episodes. Moreover, forest soils are also thought to assist in purifying the water (Chivian 2002). Forest canopies reduce the force with which rain water hits the soil and roots bind soil (Chivian 2002). Thus, forests also protect the soil against erosion and loss of nutrients. Perishing of 1400 people in the Philippines recently due to landslides initiated by rains but exacerbated by rampant illegal logging (Anonymous 2004b) illustrates the importance of rain forests for soil stability.

Medicinal and other products

Tropical forests are the source of food, remedies, natural products and construction material for many local communities. At least 25% of western

patented medicines are derived from medicinal plants identified and prepared through traditional indigenous techniques (Posey 1999). Forest products remain the only pharmaceutical option for certain remote local
communities (see Chapter 4). In addition, tropical forests remain a major
source for a variety of commercial products ranging from latex to perfumes.
There has been a recent discovery that two species of the *Calophyllum* tree
from the rain forest of Sarawak may produce anti-HIV (human immunodeficiency virus) agents (Chung 1996). The therapeutic value of rain forests
remains vastly under-explored, with myriads of new medicines and products
likely to be awaiting discovery (Laurance 1999).

Human health and ecosystem services

Destruction of rain forests can also facilitate the spread of human diseases.
Malaria is one of the main causes of sickness and death in the developing
tropics. Contemporary deforestation seems to lead to an increase in the distribution of mosquitoes and concomitant increases in mosquito-borne disease
transmission (van derr Kaay 1998; Kidson *et al.* 2000; Norris 2004).
Deforestation improves the habitat of mosquitoes by compromising drainage,
increasing light and temperature to facilitate the growth of algae (main food of
mosquito larvae) and deacidification of standing water (Chivian 2002). Several
Anopheles spp. mosquitoes are expanding in Southeast Asia, very likely to be
due to changes in land use (Aiken & Leigh 1992).

Logging activities, slash-and-burn farming, road construction and transmigration projects bring 'new' people into areas, most of whom have little or
no immunity to the tropical human diseases in the forest ecosystem (Aiken &
Leigh 1992). On the other hand, new colonists to the rain forest are also
major carriers of malaria (Moran 1988). Human mortality due to malaria
can be as high as 25% in some plantations (Ooi 1976). In the Amazon, it is
estimated that every 1% increase in deforestation boosts the number of malaria
carrying *Anopheles darlingi* by 8% (Pearson 2003). Similar increases in prevalence are expected intuitively for other insect vectors such as blackflies
(*Simulium* sp.). Increases in rat (*Rattus rattus*) abundance in oil palm plantations can also make humans more vulnerable to scrub typhus, which is transmitted by the rat's mites. The disease situation is exacerbated in new
settlements due to poor sanitary conditions. Further, these 'new' people expose
the indigenous forest-dwellers to exotic diseases that they might have brought
in. For other disease-related examples in the Southeast Asian context, see
Chapter 4.

Many forest animals, such as bees and birds, help to maintain vital
ecological processes such as pollination and seed dispersal. Nearby forested

areas are essential for the pollination of some agricultural crops (see Chapter 3). Therefore, deforestation may impact adversely the agricultural production of many areas in Southeast Asia. The loss of such 'free' ecosystem services can ironically cost dollars to humanity (Myers 1996; Ricketts *et al.* 2004). The classical example of this necessity is illustrated by the oil palm, which is native to the humid regions of Africa and was imported in the early 1900s into Southeast Asian countries such as Malaysia and Indonesia because of its commercial potential. However, the weevil (*Elaeidobius* spp.) that pollinated the oil palm in its native realm was not imported at the same time. This led the Malaysian and Indonesian planters to rely upon expensive and labour-intensive hand pollination (Syed *et al.* 1982). In the early 1980s, the most abundant and efficacious weevil pollinator (*E. kamerunicus*) was released in Malaysia (Krantz & Poinar 2004), the establishment of which boosted fruit yield by up to 60% and generated savings in labour costs worth US$140 million per year (Kevan *et al.* 1986; Chivian 2002).

Considering all of the above, it is abundantly clear that wild nature provides many benefits to humanity. Overall, it has been estimated that the conservation of natural habitats can generate at least a 100-fold greater value of economic benefits (e.g. through fuel, fiber and pharmaceuticals) than destroying and converting habitats (Balmford *et al.* 2002). In addition, there are strong aesthetic, moral and spiritual benefits that biodiversity provides for us. Human cultures and biodiversity are inextricably tied to each other, far beyond simple measures of monetary value (Gaston & Spicer 1998; Posey 1999). Many indigenous communities within Southeast Asia protect forested areas as 'worship forests' or 'sacred groves'. These worship forests are used during religious festivals and rituals, or as graveyards (Grove *et al.* 1998; Posey 1999). Yet worship forests continue to be degraded by contemporary human activities, indicating cultural erosion that has its source in deforestation. Clearly, the benefit to humanity of conserving nature has many foundations; economic, ecological, moral and sociocultural.

Loss of knowledge

The biodiversity of Southeast Asia is poorly studied compared to that of other tropical or sub-tropical regions (e.g. Central and South America, Australasia and Africa), particularly over the past 20 years (Fig. 1.11). The distribution of research effort in Southeast Asia is also noticeably biased. For example, there is an alarming dearth of information on certain taxonomic groups, such as vascular plants (see Chapter 3), which suggests that the documented biological impacts of habitat disturbance may be just

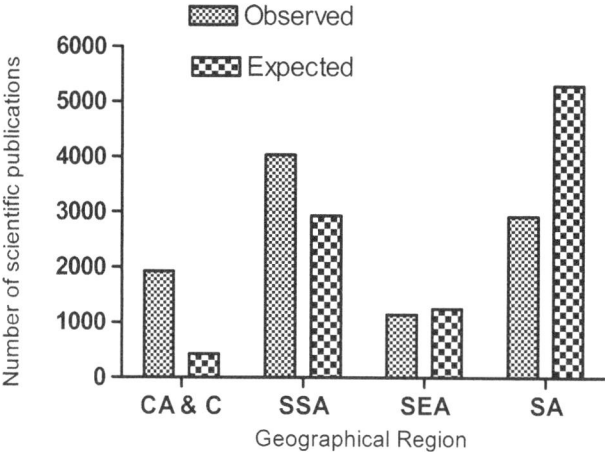

Figure 1.11 Biodiversity research effort (number of scientific publications) in tropical regions. Relative biodiversity research effort across four geographical regions, including Central America and Caribbean (CA & C), Sub-Saharan Africa (SSA), Southeast Asia (SEA) and South America (SA). All comparisons were based on the number of internationally peer reviewed research articles on biodiversity (excluding exclusively marine studies) published between 1983 and 2003 extracted from the database *Biological Abstracts*. The expected number of publications for each geographical region and taxonomic group were calculated by dividing the number of publications evenly among geographical regions and taxonomic groups, respectively.

the tip of a grossly underestimated iceberg. Undoubtedly, many great scientific discoveries are hidden in the Southeast Asian habitats. Some of these may even potentially impact humanity, such as the discovery of new medicines probably to halt the spread of emerging diseases (also see Chapter 6) Thus extinctions through habitat loss will not only impede scientific knowledge; they may also affect profoundly human well-being.

Summary

1. Habitats, particularly rain forests, are disappearing rapidly from Southeast Asia.
2. Drivers of this massive ongoing land conversion relate largely to the ever burgeoning human population and its related activities, such as agriculture, urbanisation and logging.
3. Weak institutions and rampant corruption thwart efforts to curb deforestation.

4. Habitat destruction in Southeast Asia, in addition to being a direct determinant of biodiversity loss, may elevate the emission of greenhouse gases, alter regional and local weather conditions, enhance the spread of disease, and thus ultimately affect humanity in many unpleasant ways.
5. Lack of sound scientific knowledge in the region hinders effective conservation management of habitats and the associated biota.

Chapter 2

Biodiversity in a hotspot

In this chapter we describe the uniqueness of the geographical and geological history of Southeast Asia and how these have combined to influence the high biodiversity and endemism characteristic of the region across a wide range of plant and animal taxa. Further, the biotas of Southeast Asia are compared with those of the tropical forest regions of Africa and South America, in terms of the biogeographical and evolutionary similarities and differences in their taxa.

The political entity of Southeast Asia consists of the nation states of Brunei, Cambodia, Indonesia, Laos, Malaysia, Myanmar, Philippines, Singapore, Thailand and Vietnam. However, it is worth noting that the moist tropical forests of the Western Ghats in southwest India and Sri Lanka show affinities with Southeast Asian taxa (Whitmore 1984; Morley 2000). Myers *et al.* (2000) identified 25 'biodiversity hotspots', spread across the globe, as areas containing high concentrations of endemic species and suffering substantial habitat loss. Southeast Asia overlaps with four of these hotspots (Indo-Burma; Sundaland; Wallacea; and the Philippines), each of which has a unique geological history that has contributed to its rich and often unique biotas (Mittermeier *et al.* 1999). However, the biogeography of Southeast Asia is extremely complex and not completely understood.

The first major influence on the biogeography and biodiversity pertains to the geological history of the region. The movement of tectonic plates, climate change and changes in sea-levels have, at different times in prehistory, created islands, formed land-bridges and exposed or inundated large areas of land. During the Pleistocene glacial episodes, some temperate species expanded their ranges southwards into Indo-Burma and retained their presence thereafter (Jablonski 1993). Fluctuating sea-levels periodically converted mountains into geographically isolated islands, creating conditions that were ideal for speciation. The episodic sea-level changes also repeatedly connected the islands of Sundaland to the Asian mainland, allowing for biotic migrations from the mainland (Meijaard 2004). As the sea-level rose, the isolation of these islands facilitated speciation. The presence of rain forest refugia in parts of Sundaland during the Pleistocene, a period of heightened regional aridity, also enabled the persistence of its forest biotas (Gathorne-Hardy *et al.* 2002b). Although it was never connected to the Asian mainland, Wallacea is one of the most geologically complex regions in the world, because its islands originated from land fragments that rifted from Gondwanaland at different geological time periods (Audley-Charles 1983). This unique geological history, together with its stable tropical climate and numerous insular biotas, enabled the evolution of highly endemic biotas in Wallacea. The other geologically unique region of Southeast Asia, the Philippines, consists of approximately 7000 islands, containing multiple centres of endemism (Mittermeier *et al.* 1998). The colonisation of newly formed oceanic islands, followed by genetic differentiation and long-term persistence, has resulted in the extraordinarily high species richness and endemism of the Philippines (Steppan *et al.* 2003). As a result of Southeast Asia's unique geological history, the region ranks as one of the highest in the world per unit area in terms of species richness and endemism (Mittermeier *et al.* 1998).

The second major influence on the Southeast Asian biota is the region's central position between major bioregions. The zoogeography of Southeast Asia contains the Austro-Malayan, the Indo-Chinese and Indo-Malayan sub-regions and borders the Indian, Australian, Polynesian and Manchurian sub-regions. The phytogeographic regions comprise the Malesian and Continental Southeast Asiatic and borders the North and East Australian, Melanesian and Micronesian, Sino-Japanese and Indian regions (Wallace 1876; Good 1974).

A third factor is the recent geography of Southeast Asia. Over 50% of the area of the ten Southeast Asian countries consists of islands, with two of the nations, Indonesia and the Philippines, being entirely insular. As will be discussed later, speciation is largely driven by allopatry i.e. geographical separation, of which islands are ideal examples. There is also a relationship between island size and size of the biota (Koh *et al.* 2002). Furthermore, the Southeast Asian region features unique ecological processes, such as the strong synchrony of fruiting of trees (mast events) from the Dipterocarpaceae, which has major implications for forest ecology and conservation (Curran *et al.* 1999).

All of these factors have contributed to the evolution, diversification and extinctions observed in current and historical biotas of Southeast Asia. Their relative contributions and influence on specific fauna and flora are discussed in more detail below.

Geological history

Tectonically, continental Southeast Asia is a mosaic of allochthonous terranes bounded by Siberia to the north and Kazakhstan to the west. A terrane is a crustal block or fragment that preserves a distinctive geologic history that is different from the surrounding areas and is usually bounded by faults. These terranes have histories that are variably decoupled from that of their neighbours and so their contiguity today is no guarantee of proximity in the past (Fortey & Cocks 1998; Metcalfe 1998). In this region at least eight tectono-stratigraphic terranes have been recognised (western Yunnan, Myanmar, Thailand, Peninsular Malaysia, Vietnam, Laos, Cambodia, Sumatra). At the beginning of the Mesozoic era around 250 million years ago (mya) there were two super continents, Laurasia – consisting of present-day North America, Europe and much of Asia; and Gondwanaland – containing South America, Africa, India, Australia, Antarctica and the rest of Asia. The Palaeozoic–Mesozoic framework of the tectonic evolution of these and adjacent East Asian terranes has now been broadly established, with a dominant view that most of these continental terranes had their origins in northern Gondwanaland, probably on the India–North/Northwest Australian margin. The freshwater vertebrate assemblage from the Mesozoic from Thailand is remarkable in that it closely resembles the classic vertebrate fauna from the Norian period (217–203 mya) of Central Europe, with genera in common of fishes, amphibians, turtles and tortoises indicating that dispersal of non-marine vertebrates was comparatively easy across Eurasia between Central Europe and Southeast Asia. Links with northern Asia also appear to have been established during the Jurassic (200–146 mya) as evidenced by similarities between the Southeast Asian and Chinese vertebrate assemblages (Buffetaut & Suteethorn 1998). Clearly then, the flora and fauna of Southeast Asia have had substantial inputs from the surrounding regions, and with ensuing periodic isolation there has been enhanced speciation and endemism. The geological history of the East and Southeast Asia continent can be seen as a process of step-wise accumulation of continental slivers by several episodes of intensified rifting and accretion/subduction interposed by relatively long periods of drifting. Multidisciplinary data suggest that the East and Southeast Asian terranes were successively rifted and separated from Gondwanaland as three continental slivers in the Devonian (416–359 mya), late Early Permian (299–280 mya) and late Triassic–early Jurassic (228–176 mya). The separation of these slivers of continent was accompanied by the opening (and subsequent closing) of three ocean basins (Metcalfe 1998; Shi & Archibald 1998).

What are now southern Tibet, Burma, Thailand, Peninsula Malaysia and Sumatra were once part of Gondwanaland, but rifted from the northern Australia–New Guinea continental margin about 200 mya. This dissected the land mass connection between Asia and Australia. Borneo, Sumatra and western Sulawesi and the Banda Arc islands (Flores, Alor, Wetar, Banda, Raijua, Timor, Tanimbar, Kai, Seram and Buru) are thought to have split from Gondwanaland in the middle Jurassic (176–161 mya). Eastern Indonesia is one of the most complex geological regions in the world. East Sulawesi, together with New Guinea and Australia, separated from Antarctica and moved north beginning in the early Mesozoic. At least part of east Sulawesi probably split from New Guinea before east and west Sulawesi collided, after which the eastern part emerged as an island. Sediments from the straits indicate that Borneo and Sulawesi have been separated for at least 25 mya. However, during periods of lower sea-levels it is likely islands would have existed, especially in the region of the Doangdoangan shoals and west of Majene. A drop in sea-level of 100 m would have exposed an almost continuous land-bridge between southwest Sulawesi and southeast Borneo (Whitten *et al.* 1987a).

It has been suggested that eastern Borneo had collided with western Sulawesi, thereby closing the Makassar Straits in the late Pliocene (3 mya) (Katili 1978). However, others believed that there was not a great deal of evidence for this (Whitten *et al.* 1987a). Movements of continents and continental fragments and the development and destruction of oceanic basins during the evolution of the Southeast Asian region have thus resulted in the creation and destruction of biogeographic barriers at various times, in turn fostering the waxing and waning of different faunal and floral provinces. In the mid Eocene the collision of the Indian and Southeast Asian plates permitted the merging of their respective biotas. Since the Indian plate carried elements of the African Cretaceous flora, plants of African origin arrived as well as the Indian endemics. Secondly, the collision of the Sunda and Australian plates created pathways for the movement of Gondwanaland species into Southeast Asia. Thirdly, the presence in East Asia of a continuous continental connection across latitudinal climate zones from the equator to 60° N allowed the survival of diverse elements from the northern hemisphere Boreotropical Province in this region in a manner not seen elsewhere. At continental margins, such as on the Sunda and Sahul shelves, changes in climate and sea-level may have led to patterns that reflect both the migration of ecological zones and the disruptive influence of marine transgression. Thus, the dispersal and evolution of faunas and floras of Southeast Asia are intimately linked with the geological evolution of the region, though the fossil evidence for the history of the Southeast Asian region is varied (Steppan *et al.* 2003). In contrast to Southeast Asia, with its abundance of islands, South America has comprised a single land mass since its split from Gondwanaland with periods of connection and isolation from North America. Africa was also largely one plate. The African lithographic plate bears the largest of the three tropical land masses. It has also been the most geographically isolated, and stable throughout much of its Tertiary history, and is currently the most elevated (Morley 2000).

Climate change

The history of the flora and fauna of Southeast Asia is also tied closely to climatic changes. Climate, in particular variation in solar energy input and evapo-transpiration, are seen as major factors promoting the high levels of biodiversity at low altitudes. In drier periods, seasonal forests and savannah would have expanded as areas of rain forest contracted. At least some rain forest species would have remained in isolated pockets especially along riparian habitats and moister soils, thus some level of gene flow could be maintained (Currie 1991; Wilson & Rosen 1998). The richness of the present Southeast Asian flora is due at least in part to the widespread dispersal of megatherms (plants requiring continuous high temperatures with abundant precipitation) in the region, and the mixing of floras from different geographical origins, which did not occur to the same extent in Africa and the Neotropics (Morley 2000).

Climate change at the end of the Eocene and Neogene

The terminal cooling of the Eocene around 35 mya appears to have been less marked in the Neotropics than anywhere else, with a reduced number of extinctions and the continuation of mauritiod palms into the Neogene (Miocene and Pliocene, 23–1.8 mya) which became extinct in Africa at the end of the Eocene. However, there was a profound effect on the vegetation in the Proto-Indian Province, initially in Southeast Asia. There was global cooling and possible changes in atmospheric circulation patterns resulting from the Himalayan uplift producing a warmer and moister climate in the Himalayan foothills, allowing the development of evergreen tropical forests which spread up from Southeast Asia. The flora of the Malesian Province (the latest early Miocene and later flora up to the present day) underwent further changes with the collision of the Sunda and Australian plates and the Miocene formation of the islands of eastern Indonesia and New Guinea. Montane floras became particularly enriched by dispersals from the Australian plate into the East Malesian floristic province. As sea-levels fell, islands arose and land-bridges opened. This helps to explain the predominance of Malesian elements in the islands of New Guinea and eastern Indonesia. There was a stranding in southwest

Sulawesi of a fragment of the Sundanese Eocene flora and this provided a source for more aggressive Malesian elements to colonise from the west. Palynological data from the Southeast Asian Neogene also demonstrates that the diversity of the Southeast Asian flora became accentuated as a result of the successive re-formation of lowland vegetation on the continental shelves during periods of low sea-levels. Diversification of tropical rain forest flora continued through a major part of the Tertiary as a result of the successive expansion and retraction; and fragmentation by physical barriers (Holloway & Hall 1998; Morley 2000).

Climate change in the Quaternary: glaciations, interglacials and refugia

The climate of the region today is quite different from what predominated during and before the Quaternary. Tropical and sub-tropical conditions and their associated biotas extended, from the equator, further north and south than they do now. In the late Quaternary temperatures repeatedly rose and fell. In periods of lower temperatures the ice caps at the poles expanded, binding up vast quantities of seawater, causing the lowering of sea-levels worldwide. The maximum drop in sea-level (Holloway & Hall 1998) in Southeast Asia was around 150–180 m (Whitten *et al.* 1987a; Whitten *et al.* 1987b). This fall in sea-level exposed vast areas of land, and what were once islands became connected or re-connected to other land masses. During the Pleistocene epoch of the Quaternary period (1.8 million to 11 400 years ago), there were several climate changes; warm wet pluvial periods alternating with cold dry interpluvials, correlating with periods of more intense glaciation in temperate regions and corresponding lowering of sea-levels of 100 m or more as water was locked up as ice at the poles. In Southeast Asia, this repeated pattern of glaciations induced greater climatic seasonality. This caused a change in the botanical composition from rain forest to savannah or a wooded savannah flora, with some rain forest refugia. The lowering of sea-levels resulted in the exposure of vast areas of land, which are presently submerged. Borneo, Sumatra and Java were all connected to mainland Asia by land-bridges, with the last connection only 10 000 yrs ago (Morley 2000). The Sunda shelf, where Bali, Borneo, Java and Sumatra connect to mainland Asia was three-times its present area of dry land. At a drop in sea-level of 40 m there would have been direct land-bridges between Singapore and Kalimantan; Vietnam and Sarawak; Java and southern Kalimantan; and New Guinea and Australia. The climate of the main Sundanese land masses would have been one of lower precipitation and humidity and an increased diurnal temperature range. The cooling effects lowered the elevation of tree line, and also lowered the upper limits for montane vegetation (Whitten *et al.* 1987b).

Java, a smaller and more isolated land mass, was connected to Asia less often and for shorter periods than either Sumatra or Borneo. This helps to account for some of the differences in the Javanese fauna e.g. the craniometric differences between Javanese leopards and leopards from the rest of Asia. Java had at least seven species of elephants in the early Pleistocene, and numerous species that now only survive elsewhere, including orangutans (*Pongo* sp.), Malay sun bear (*Helarctos malayanus*), Malayan tapir (*Tapirus indicus*), tiger (*Panthera tigris*) and clouded leopard (*Neofelis nebulosa*). Borneo and Sumatra could not support leopards due to the low ungulate biomass and competition from other carnivores better adapted to tropical evergreen forest (Meijaard 2004), though Whitten *et al.* (1987a) had earlier questioned this reasoning for the absence of leopards, since leopards readily take monkeys and other vertebrate prey and are not restricted to large ungulates.

At the Last Glacial Maxima (LGM), some 18 000 years before present, there was an approximate fall in temperature of 4 °C. This would have dropped the altitudinal range of montane vegetation zones by about 700 m. There is some palynological evidence of montane forest migrating downhill in cooler times in Sumatra and Java (Newsome & Flenley 1988; Stuijts 1993). During periods of intense polar glaciation, there was a reduction in rainfall and temperature in Southeast Asia, but lowland rain forest could continue to survive thanks to its proximity to the sea due to the presence of convergent precipitation with oceanic moisture. There are also indications of the presence of continuous lowland rain forest by the inter-riverian endemism of dipterocarps. Much of the ice-age vegetation was a mixture of savannah woodlands and intermittent deciduous forest. During the LGM most of Thailand, Peninsular Malaysia, western and southern Borneo, eastern and southern Sumatra, and Java were probably covered by savannah vegetation types. Only pockets of rain forest remained in a few refugia in northern and eastern Borneo, northern and western Sumatra and the Mentawai Islands (Brandon-Jones 1998; Gathorne-Hardy *et al.* 2002b; Meijaard 2003). Forest obligates were driven to the refugia, leading to the proliferation of savannah vertebrate species such as mastodons, elephants, stegodonts, antelopes, hippopotamuses and bovids (van den Bergh *et al.* 2001). During the interglacials, rain forest species were able to recolonise and savannah animals were constrained to Indo-China and a few patches of Java. Few attempts have been made to delineate forest refugia for Malesia, due probably to its complex patterns of islands, but forest survival at intermediate altitudes on volcanic islands seems certain, even if it disappeared from parts of Borneo and from northern Queensland due to large-scale variations in rainfall patterns and volumes. Over the humid tropical core areas the changes are likely to have been less severe, but the extent of these areas is unclear (Kershaw 1978; Thomas 2000). High rates of non-volant mammal endemism in the Mentawai Islands and present day primate distributions indicate the continued presence of lowland forest refugia (Brandon-Jones 1998; Morley 2000; Gathorne-Hardy *et al.* 2002b).

Each glacial episode probably showed similar patterns of invasions, extinctions, isolations and speciation. There is further evidence of climate change and the continued presence of rain forest refugia from studies of extant termite species. Southeast Asian termites are generally poor in terms of numbers of species (with the exception of the *Termes/Capritermes* group) if one compares them to the Americas and Africa. Termites are sensitive to disturbance, especially to forest canopy loss, their evolutionary recovery is slow, and they are generally poor at crossing zoogeographic barriers. The composition of termite species provides an indication of historical patterns of stability, making them ideal organisms to investigate the effects of Quaternary climate change (Bowman 2002). For speciation to occur in soil-feeding termites, a large area of lowland rain forest has to remain stable for long periods. A comparison of the termite species at supposed savannah sites and refugia showed different species compositions indicating different prehistories for the sites (Gathorne-Hardy *et al.* 2002b)

Southeast Asia, South America and Africa show major differences in the representation of upland and lowland terrain, which need to be taken into account when considering the distribution of tropical rain forests at the LGM. The worldwide cooling of tropical climates at the LGM now seems almost indisputable, and according to some recent findings the magnitude of temperature variation was similar at high and low altitudes, with montane taxa penetrating into the lowland forests in Africa and the Neotropics (Maley 1987; Servant *et al.* 1993). The argument is that there was a general shift of forest downwards to lower elevations as temperatures decreased. Tropical species became extinct when their thermal minima were breached leaving niches for more species adapted to lower temperatures.

It is clear that Africa experienced the driest climates at this time, but also, with easily the lowest representation of low altitude terrain and minimal representation of continental shelves, provided by far the fewest opportunities for rain forests to find refuge in the lowlands (Morley 2000).

In South America there is some evidence of drying during the LGM, though not as pronounced as in Africa. In South America, during cooler phases of the Pleistocene, lower temperatures pushed lowland rain forests to the lowest altitudes, and at mid-elevations lower montane taxa grew together with lowland rain forest elements in plant associations that have no modern analogue, maintained by orographic rain (rainfall produced by supersaturated air mass forced upwards by a mountain range). Thus the world's most species-rich rain forests along the lower flanks of the Andes probably correlate to continued orographic rain over a long time period, through both glacial and inter-glacial periods, whereas elsewhere the distribution of lowland rain forests was restricted by a combination of lower temperatures and reduced moisture. The desiccation associated with glacial maxima does not seem to have been so pronounced in South America as in Africa and these moist conditions probably enabled the survival of very diverse rain forest communities through the most unfavourable periods of the Pleistocene (Morley 2000).

In Southeast Asia, dry climates clearly affected large areas, but the huge land areas which would have become available when sea-levels were more than 100 m below their present levels, would have compensated significantly for the expansion of drier climates. In South America, the expansion of dry climates was probably much less marked than in Africa, and diverse rain forest floras found refuge by migrating to the river basin floors; however, as for Africa, little terrain became available with lower sea-levels, due to the steep continental slope. During the Holocene about 60% of the area of the African continent lay within the altitudinal range of tropical rain forest compared to 70% of South America and Asia. During the LGM with the elevational extent of tropical forests possibly depressed by 500 m, 40% of the area above present sea-levels in Asia and 35% in South America was within the altitudinal range of tropical rain forest, but in the more elevated African continent, only 15% of the land area would have been within this altitudinal range. The limited representation of low altitude terrain should be given as much attention as moisture reduction in attempting to explain diversity loss in Africa (Harrison et al. 1981; Morley 2000).

Thus, it is most likely that the richest centres of species-diversity and endemism in both South America and Southeast Asia are the result of long-term continuity of moist climates within the equatorial zone. This makes earlier ideas that species-richness was a reflection of long-term stability more understandable. The alternative suggestion that areas of highest species richness are not areas of maximum stability but of maximum disturbance (Bush 1994), requires further critical examination.

Pleistocene refuge theory and the 'species pump'

As mentioned above, during the ice-ages there were contractions of tropical rain forest in Asia, Africa and the Neotropics into refugia. These successive isolations of populations restricted gene flow, giving rise to the idea that these Pleistocene refugia acted as a 'species pump' driving speciation (Haffer 1969, 1987). However, it has been argued that Amazonian forests did not fragment, but populations were affected by changes in climate, with areas of high species richness being due to increased disturbance (Bush 1994). There is a lack of evidence for late Pleistocene refugia for South American mammals (Lessa *et al.* 2003) and Morley (2000) believes that it is the long term continuity of humid tropical rain forests in Southeast Asia and South America that has produced the highest diversity and endemism, rather than the 'species pump'. It is likely that speciation has taken place over a much longer time, with different forcing mechanisms (e.g. from orogeny (the process of mountain formation) and the formation of large rivers) inhibiting dispersal.

However, the 'Pleistocene refuge theory' helps to explain many of the plant and animal distributions seen in African rain forests, which were reduced to a fraction of their present distribution as dry climate vegetation expanded during glacial maxima. The African rain forests have been described as the 'odd man out' on account of their much lower species-richness and the poor representation of important groups which are well represented elsewhere (Richards 1973). The whole of continental Africa has fewer palms than Southeast Asia's small island of Singapore, orchids are poorly represented, and there is only a single species of bamboo. In contrast, there are 41 indigenous species of bamboo in Thailand, while Malaysia has 25 and Indonesia has 35. Discontinuities in the geographical distributions of African birds and rain forest plants are believed to reflect the restriction of rain forest to refugia during the driest and coolest periods of the Pleistocene, from which they subsequently spread during the Holocene. There appears to be no record of Middle Miocene (16–12 mya) floristic interchange with Eurasia, comparable to that seen in mammals. This was probably due to the marked zonality of climate at this time, rendering long-distance dispersal across bioclimes unlikely. However, it is possible that temperate taxa were able to disperse along the highlands of East Africa and Eurasian warm temperate plants were able to extend into North Africa. Following global climate change at the end of the Middle Miocene, savannah successively expanded over much of the African tropics at the expense of rain forests. At the same time large areas of the continent underwent uplift, resulting in more limited opportunities for the dispersal of lowland rain forest species into refugia since there was only a restricted presence of suitable low altitude habitats. This resulted in widespread extinctions within the rain forest flora from the late Miocene into the Pliocene (Morley 2000).

Land-bridges, islands, isolations and sea-level changes

A substantial proportion of the land area of Southeast Asia consists of islands and island groups. The number of islands, both in the aquatic sense and in the sense of one habitat surrounded by a different habitat forming a barrier that makes crossing difficult, is unique to Southeast Asia. However, there are transitional zones with influences from neighbouring faunal zones. Different biotic groups (and even among members of the same genera) have differing dispersal abilities across water and other potential barriers. Furthermore, periodic episodes of rises and falls in sea-levels have at various times isolated or permitted connections, allowing freer movement of species. These episodes of isolation have driven speciation and endemism.

There have been studies of speciation and biogeography in extant small mammals e.g. murine rodents in the Philippines (Steppan *et al.* 2003) and the Sunda shelf (Gorog *et al.* 2004), and gymnures (Insectivora) (Ruedi & Fumagalli 1996). None of these studies supports the idea of speciation in these mammals being strongly driven by migration over Pleistocene land-bridges. The murine research of Gorog *et al.* (2004), showed that there were closer affinities between the Malayan Peninsula and Greater Sundanese island lineages than between the Malayan Peninsula and Indo-Chinese groups. Ruedi & Fumagalli (1996) described a similar pattern and proposed a sequential dispersal model, where the mainland source area is represented by the most pleisiomorphic lineage (i.e. ancestral) and more recently colonised islands are represented by more recently derived lineages. However, Gorog *et al.* (2004) argued that the observation of this pattern in a number of unrelated taxa is perhaps more plausibly due to a long history of vicariant evolution affecting the Malay Peninsula and then the Sunda Islands. This may correspond with the Pliocene fragmentation of the Sunda block.

There are two principal assemblages of animals in South and Southeast Asia which are based on the Sunda and Sahul continental shelves of Asia and Australia, respectively. Intervening is a transitional fauna of Sulawesi and associated islands. A rather special assemblage exists on the Mentawai Islands which are separated from Sumatra and the Sunda shelf by a deep ocean trench. The Greater Sundas can be divided into three faunal subgroups: (1) the Sahul shelf of the Aru Islands, New Britain and New Guinea with its Australo–Papuan elements; (2) the Sundaic sub-region of Borneo, Sumatra, Java and the Malaya Peninsula (faunistically more similar to the Sundas than to the rest of mainland Asia); and (3) Wallacea, the Philippines, Moluccas, Sulawesi and the Lesser Sundas. These islands lie between the two major shelves, the Sundaic and the Sahul. We will look as the assemblages of animals in each of these areas separately.

The Sahul shelf of the Aru Islands, New Britain and New Guinea with its Australo–Papuan elements

Even though most birds are highly mobile there is a sedentary component of an avifauna. The proportion that is sedentary depends on several factors, including distance to areas that have the potential for colonisation. It is thought that 70% 'sedentary species' was general for avifaunas in the Indo-Pacific, based on work in New Guinea/New Britain, and 30% were able colonisers (Mayr 1941). Many of the New Guinean mammal groups have similarities with Australian taxa, in particular those species from Cape York and the Queensland coast. Monotremes and marsupials are present in New Guinea and two genera of marsupials (phalangers) reach Sulawesi. The north and central Moluccas had some endemic marsupials, most of which are now extinct (Flannery 1996).

The Sundaic sub-region of Borneo, Sumatra, Java and the Malayan Peninsula

Borneo, Sumatra, Java and the Malay Peninsula were isolated from each other some time between 9500 and 7000 years ago. If large scale migrations occurred periodically up until this time, the subsequent period of isolation is unlikely to have resulted in complete lineage sorting and pronounced genetic differentiation among populations in these regions. There is still some debate as to the extent of rain forest in the Sunda region at this time. Some studies demonstrate a predominance of grassland plants character-istic of savannah or steppe vegetation (Morley 1998; Morley 2000), whereas others indicate that tropical rain forest and mangrove swamp dominated at least the core southern region of the shelf (Sun et al. 2000; Kershaw et al. 2001). The movements of rain forest taxa may have been facilitated by the large rivers that dissected the Sunda region during this period. Today, beneath the South China and Javan Seas lie deep troughs that formerly connected extant Bornean rivers to extant rivers in Java and Sumatra. Other rivers flowed between Sumatra and the Malay Peninsula (Voris 2000). These rivers enabled freshwater fish to move among them and, if the riverine systems supported gallery forests, they may have also acted as dispersal corridors for rain forest organisms (Dodson et al. 1995).

There is often a rapid turnover of taxa at provincial boundaries. Good (1974) discussed the differences between the zoogeographical transition from the Indo-Chinese sub-region to the Malayan sub-region and the cor-responding phytogeographical transition between the continental Southeast Asiatic Floristic Province and the Malayan Floristic Province of the Indo–Malayan Subkingdom. This has been described as the transition between the Indo-Chinese Bioregion and the Sunda shelf and Philippines Bioregion (Wikramanayake *et al.* 2002). Although the position of the phytogeographic transitions, involving over 500 genera, can be determined today by climatic factors, with the avifaunal transition probably linked causally (directly or indirectly) to the flora, it still does not explain fully the origins of the different Indo-Chinese and Sundaic biotas.

Within most of the bird families some species appear to be confined to either northern Indo-Chinese or southern Sundaic provinces, while others occur across the whole latitudinal range. About 190 species are exclusively Indo-Chinese, 150 are Sundaic and another 147 species are widespread. Some 56 species are 'montane island-hoppers' occurring in the hills of northern Thailand and those of Peninsular Malaysia, but not in the intervening central Thai–Malay Peninsula. Examples of families that best illustrate these patterns of species replacement with congeneric Indo-Chinese and Sundaic species include the woodpeckers (Picidae), pheasants (Phasianidae) and bulbuls (Pycnonotidae). In the woodpeckers there are 11 Indo-Chinese species, 16 Sundaic species and eight widespread species of which five show significant range gaps in the northern and central peninsula. In the pheasants, there are 11 Indo-Chinese, seven Sundaic and three widespread species. In the bulbuls, 11 Indo-Chinese species are replaced by 17 Sundaic species with seven more species being widespread. With both the woodpeckers and bulbuls, relatively more sub-speciation occurs in the southern than in the northern species.

Thus at the species level the Sundaic avifauna is clearly different from that of the Indo-Chinese sub-region. Overall the pattern is of a much richer lowland forest fauna in the Sundaic than the Indo-Chinese province and perhaps this is linked to the greater structural complexity of the less seasonal rain forests. Many Sundaic species have no ecological equivalents in the Indo-Chinese fauna. In other cases one or two Indo-Chinese generalists are either replaced, or added to, by further congeners in the Sundaic province. Among leafbirds (*Chloropsis* spp.), for example, the golden-fronted leafbird (*C. aurifrons*) is found exclusively north of the Kra Isthmus within the region covered (although it also occurs in north Sumatra): the blue-winged leafbird (*C. cochinchinensis*), occurs on both sides of the divide, as does a montane species, the orange-bellied leafbird (*C. hardwick*). Two other lowland species, the greater green leafbird (*C. sonnerati*) and the lesser green leafbird (*C. cyanopogon*), are exclusively Sundaic (Hughes *et al.* 2003).

There is a marked avifaunal transition north of the Isthmus of Kra (the narrow neck of land linking the Malay Peninsula to the rest of the Asian continent) and no detectable cluster of species boundaries further south near the transition between perhumid and wet seasonal evergreen forest types. These analyses support the hypothesis that a significant turnover in bird species assemblages occurs between 11° and 13° N on the Thai side of the Thai–Malay peninsula. The position of transitions may be associated with present or former barriers to dispersal. The significant transitions on the Thai–Malay peninsula require that some sort of barrier existed in the past in order to permit the provincial biotas to diverge in allopatry (Woodruff 2003). Unlike the Central American biogeographic transition or the Asian–Australasian transition associated with Wallace's Line, this avifaunal transition near the Isthmus of Kra cannot be attributed to plate tectonics and former open marine barriers to dispersal. Although the timings of sea-level changes in the Pleistocene are not well understood there is evidence that there were periodic inundations around this isthmus (Rangin *et al.* 1990; Ridder-Numan 1998). Woodruff (2003) suggested that the Isthmus of Kra was inundated by the sea at two different areas and this helps to account for the current biogeographic boundaries and patterns. For a number of taxa, lowland birds (Wells 1971; Medway & Wells 1976), amphibians (Inger 1966) and murid rodents (Musser & Newcomb 1983; Lekagul & McNeely 1988), the isthmus represents a peak in species limits, separating Indo-Chinese taxa from Sundanese. Amphibian species also show turnover with the isthmus but more Indo-Chinese species extend south than Sundaic species extend north of the transition (Inger 1966). Interestingly, the opposite tendency was reported in butterflies (Corbet & Pendlebury 1992). The geographical ranges of mammals showed that, like the avifauna, the boundaries of numerous mammalian species and sub-species lie near the isthmus (Lekagul & McNeely 1988; Corbet & Hill 1992). An analysis of the extant rodent fauna found that, of 61 species, 61% have range limitations associated with the isthmus (Chaimanee 1999). There is a wide area around the isthmus where the two Southeast Asian continental sub-species of macaque monkeys intergrade (Groves 2001). Within the carnivores, dhole (*Cuon alpinus*), clouded leopard (*Neofelis nebulosa*) and the yellow-throated marten (*Martes flavigula*) all have sub-species in this region separated by the Isthmus of Kra (Kanchanasakha *et al.* 1998).

In mainland Southeast Asia (extending from Myanmar in the west through to Vietnam in the east and the tip of the Malayan Peninsula in the south), there are 30 genera and 48 known species of carnivores. In the northwest of the region, the Myanmar–Thai–Indochina Peninsula forms an ample connection to the heartland of Asia. On a geological time-scale there has been no impediment to invasion and gene exchange between wild populations. Carnivores are usually highly mobile and therefore have a natural tendency to disperse easily over large areas. Some 300 species of mammal and around 1000 species of birds have been recorded from

Myanmar. This incredible diversity is due to the country straddling two major zoogeographic areas. The south has a strong Malesian influence, the central and north have Chinese and Indian affinities while a large number of Himalayan species occur in the north and west. A total of 68 species of swallowtail butterflies (12% of the world's known species) have been recorded. Only Indonesia, China, Brazil and India exceed this number of species. Eight genera are present in Myanmar, *Bhutanitis, Meandrusa, Lamproptera, Graphium, Atrophaneura, Troides, Papilio* and the designated rare status, *Teinopalpus imperialis* (Collins & Morris 1985).

Overall, the avifauna contains 12% of known global bird species, but shows extremely low endemism (Smythies 1953).

Many of the mainland Southeast Asian species also have ranges that extend into Sumatra, Java and Borneo. Borneo, for example, has 26 species of carnivore but only two are endemic, Hose's civet (*Hemigalus hosei*) and Hose's mongoose (*Herpestes hosei*), known only from a single specimen collected in 1893 (Payne *et al.* 1985). Borneo has the larger land mass but is more distant from Asia than Sumatra so as a consequence the avifauna is slightly less species rich than Sumatra's but has more endemics. Most of these are montane since during the periods of intense glaciation, drier low-lands were re-colonised from Asia and the mountains remained as evergreen forested 'islands'. A number of carnivore species have Indo-Chinese ranges with congenerics in Malaysia and into Indonesia, particularly in Sumatra e.g. Tonkin otter-civet (*Cynogale lowei*) and the Sunda otter-civet (*C. bennettii*). Others are distributed throughout the region e.g. Asian golden cat (*Catopuma temmincki*), masked palm civet (*Paguma larvata*) and the smooth-coated otter (*Lutrogale perspicilata*). Others have disjunct distributions, for example, the Eurasian otter (*Lutra lutra*) is recorded in Myanmar, northwestern Thailand, northern Laos and the northern half of Vietnam, is absent from Cambodia, most of Thailand and Peninsular Malaysia, but reappears in Sumatra and Java (Kanchanasakha *et al.* 1998).

Java has a less species-rich fauna but a higher proportion of island endemics. Java also has a drier climate which is thus more similar to that which occurred during the maximal land-connection, so some 30 species of seasonal forest birds from mainland Asia have continued to survive there, such as the green peafowl (*Pavo muticus*), lineated barbet (*Megalaima lineata*), small minivet (*Pericrocotus cinnamomeus*) whilst presumably becoming extinct in Borneo and Sumatra (MacKinnon & Phillipps 1993).

Indonesia, with 411 species, has the world's highest number of 'restricted range species' of birds, although in terms of continents South America has the highest, with over 700 species. A restricted range species has been defined as a species with a historical, total global breeding range of less than 50 000 km^2. About 25% of the world's birds are restricted range species and forest is their most important habitat type (Long *et al.* 1996).

The Cambodia/Vietnam/Laos area is a site of high primate endemism, with four species of gibbons (*Hylobates concolor*, *H. leucogenys*, *H. siki* and *H. gabriellae*) and seven species of langur monkeys (*Trachypithecus hatinhensis*, *T. laotum*, *T. delacouri*, *T. ebenus*, *Pygathrix nemaeus*, *P. cinerea* and *P. nigripes*). The genus *Pygathrix*, the douc langurs, is endemic to this region (Groves 2001).

During the most LGM, sea-levels fell and potential land connections arose that facilitated migration (Fig. 2.1). Several apparent anomalies become clear when one considers the distribution of land masses during the late Pleistocene: the marked differences between Bornean and Sulawesi faunas, despite being comparatively close to each other; why Palawan's avifauna is as closely related to the Philippines as it is to Borneo; and the similarities in faunas of southeastern Borneo to Java.

Mentawai Islands

The Mentawai Strait, a deep trench between the Mentawai Islands and western Sumatra, perhaps explains why the avifauna of Mentawai is more distinctive than of the other Mentaur Islands (which were connected to Sumatra at various times). Endemism in Mentawai mammals is marked with ten endemic species including the Mentawai macaque (*Macaca pagensis*) in two so considerably different sub-species that Groves (2001) considers perhaps separate specific status is warranted; the Mentawai langurs (*Presbytis potenziani potenziani* and *Presbytis potenziani siberu*); pig-tailed langurs (*Simias concolor concolor* and *Simias concolor siberu*); and Kloss gibbon (*Hylobates klossi*). The genus *Simias* is endemic to the Mentawai Islands. The macaque monkeys have African origins with the extant Barbary macaque (*Macaca sylvanus*) surviving in the ancestral range of North Africa. All the other *Macaca* species are in Asia with 12 species in Southeast Asia, three in Sri Lanka/Western Ghats and four in the rest of Asia from Afghanistan through to Japan. The most widespread species in the region is *Macaca fascicularis*. There are ten sub-species (if one includes the Nicobar Islands sub-species) of which at least six and perhaps eight are confined to islands (Groves 2001).

Wallacea: Malaku, Sulawesi and the Lesser Sundas

Wallacea, named after the nineteenth century naturalist Alfred Russel Wallace is a group of islands that comprise a zoogeographic transition between the Oriental and Australian regions. There is a very high number of endemic birds in Wallacea and almost all islands of significant size have at least one endemic species (Coates & Bishop 1997). Wallace's Line marks the

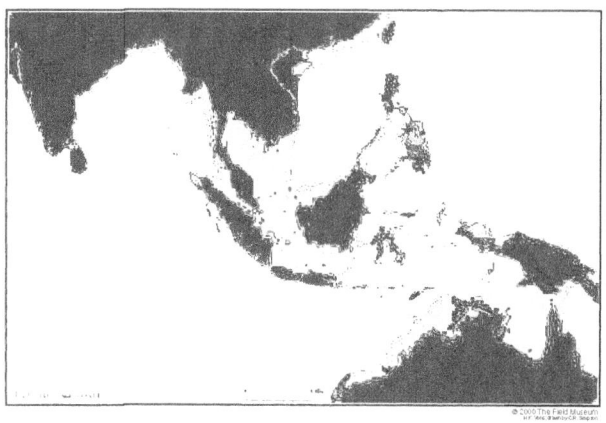

Figure 2.1 Southeast Asia and the land exposed by a 120 m fall in sea-level. Present day bathymetric depth contour as indicative of previous shore lines. (From Voris 2000. Copyright 2000 Field Museum of Natural History, Chicago.)

western limit of Australasian mammals and the eastern limit of the major oriental fauna.

In Sulawesi, of the non-endemic birds 36 of 140 species are thought to be of Asian origin. Some 25 of these species have extended from mainland Asia and are at the eastern extremes of their distributions. A further 11 species extend to the Sula islands and/or the Lesser Sundas. A total of 25 species are Australasian, with Sulawesi at the western extremity of their ranges. A further 26 species are endemic to Wallacea and the remaining 53 species cross Wallace's Line. There is about 60% similarity with the avifauna of the Philippines. The Greater Sundas (particularly Java, western Sumatra and northern Borneo), the Lesser Sundas and the southern Malaku also share species with Sulawesi. The northern Malaku avifauna shows similarities with New Guinea whilst the south Malaku avifauna shows a relationship to areas of southern New Guinea and Australia e.g. *Eos* and *Lorius* parrots. A second pattern in the south Malaku distributions links south Malaku to Asia with the presence of bulbuls, babblers, orioles amongst others (Michaux 1998).

Sulawesi has one of the most distinctive mammalian faunas, with 62% endemism (98% if one excludes the order Chiroptera, the bats) (Whitten *et al.* 1987a). The true figure is likely to be higher as only four species of *Macaca* and one species of *Tarsius* are listed in Whitten *et al.* (1987a). Groves (2001) has described six species of macaque in Sulawesi and at least three *Tarsius* species.

Although we have focused on some of the more conspicuous and hence better studied groups i.e. mammals and birds, there are also interesting biogeographical patterns in other taxa. The islands of Halmahera (northern Malaku) and Seram (southern Malaku), although now relatively proximate, have had quite different geological histories, however, this does not reflect in their respective butterfly faunas. The similarity in species composition has been attributed to the same species colonising from Sulawesi and New Guinea. Endemism is high at around 21% of the 385 recorded species. The exchange of endemic species between northern and southern Malaku has been low at less than 155 of the 82 endemic species. This indicates that the similarity is not due to exchange of species but rather independent colonisations by the same species (de Jong 1998).

Within Sulawesi parapatric speciation has been shown for a wide range of animals, although perhaps best illustrated by the macaques (*Macaca* spp.) (Whitten *et al.* 1987a; Groves 2001) a number of invertebrate groups also suggest speciation *in situ*: the *Chitaura* genus of grasshoppers (Butlin *et al.* 1998); pond-skaters, cicadas, carpenter bees, tiger beetles and limacodid moths (Knight & Holloway 1990; Cassola 1996).

The Philippines

The Philippines comprise over 7000 islands, originally entirely covered in forests of a number of types (lowland and montane monsoon; lowland and montane rain; lowland evergreen; pine; and mangrove). Much of the forest has now been destroyed, including the complete deforestation of entire islands. There is substantial endemism in both plants and animals. A total of 70 out of 96 non-volant mammals are endemic, including large species such as the bovid, tamaraw (*Bubalus mindorensis*) and two species of deer (*Cervus alfredi* and *C. calamianensis*). Of around 12 000 species of plants, 3500 are endemic and 33 endemic genera (Cox 1991).

Faunal endemism in Southeast Asia

One key factor relating to endemism and conservation in Southeast Asia is the highly insular geography of the region, coupled with the rise and fall of land-bridges in the past. In the Sundaic region there is only 0.7% endemism in the terrestrial resident avifauna. This contrasts with 45% endemism in the Philippines and 33% endemism in Sulawesi (MacKinnon & Phillipps 1993).

Although there is substantial endemism in species composition in the Southeast Asian avifauna, there are no avian families exclusively Southeast Asian. However, some account should be taken of 'Southeast Asia' being a political rather than a biological region. Some bird families are wholly Asian and well represented in Southeast Asia, but have distributions that extend to India or China as well.

Within the Indonesian archipelago, the modern and fossil land vertebrate faunas can be split into balanced mainland-influenced faunas and unbalanced endemic island faunas. Unbalanced island faunas are characterised by low numbers of genera of large herbivores and the absence of large carnivore genera. Sulawesi and the Lesser Sundas are examples of these unbalanced endemic faunas, having always been isolated from mainland Asia (van den Bergh *et al.* 2001), while the recent faunas of Kalimantan, Sumatra, Java and Bali show many components of mainland Asian fauna.

Of the 19 orders of mammals (excluding Cetacea, the whales) 15 are present in Southeast Asia the highest proportion for a tropical region (14 in Africa; 11 in South America) including two endemic orders the Dermoptera (colugos) and the Scandentia (treeshrews, although one species is present in India). Both of these orders are confined to tropical South Asian forests. Southeast Asia shares two orders of mammals in common with Africa, Pholidota (pangolins) and Proboscoidea (elephants). Both these orders are reduced in numbers of species compared to prehistoric times (Whitten *et al.* 1987b; Brook & Bowman 2004).

The primates of the Southeast Asian region exhibit a high degree of endemism. Of the 68 described species, 54 are endemic (79%) and 31 are island endemic species (46%) (Table 2.1). Overall, 57% of endemic

Table 2.1 *Primates and endemism in Southeast Asia*

Family	Endemics	Island Endemics	Total no. of species present
Loridae	2	0	3
Family **Tarsiidae**	7+	7+	7+
Cercopithecidae, Subfamily Cercopithecinae Tribe Papionini	9	7	12
Subfamily Colobinae, Langur Group	18	9	24
Odd-nosed Group	6	2	6
Hylobatidae	11	4	14
Hominidae, Subfamily **Ponginae**	2	2	2
	54 (79%)	31 (46%)	68

Taxonomy and distribution from Groves (2001). (*Homo sapiens* and sub-specific endemism have been excluded.) In bold are endemic taxa (however, the Hylobatidae do extend into southern China including Hainan Island, Bangladesh and northeastern India).

primates in Southeast Asia are island species, emphasising the link between insularity and endemism in the region. The true proportion is undoubtedly higher, since Groves (2001) believes that *Tarsius spectrum* should be split into two or three species and another *Tarsius* known from only three specimens has not yet been assigned to species level. The Family Tarsiidae, genus *Tarsius* (tarsiers); *Nasalis larvatus*, the proboscis monkey; *Simias concolor*, the pig-tailed langur; and the genus *Pongo* (orangutans) are primate genera that occur nowhere else outside the Southeast Asian islands. Furthermore, all of the 14 gibbon species, family Hylobatidae, occur in Southeast Asia, although three of these species have ranges extending into India, Bangladesh or China (and one of these is confined to Hainan Island and Vietnam only).

There are two crucial differences between the endemism characteristic of Southeast Asia compared to that of the Americas and Africa.

1. Africa and the Neotropics are, respectively, single, continuous land masses (with differing biomes). Both are essentially single tectonic plates and so their ancestral biotas spread largely by their own methods of dispersal. Throughout the Tertiary, South America, had most of its surface subaerially (on or just below the surface of the soil) exposed (currently the Amazonian Basin lies at less than 200 m above sea-level). The history of both vegetation and continental connections in Central America is crucial when assessing the

north–south Paleogene dispersal and in understanding the environment within which vertebrate migrations took place during the late
Neogene, the so-called 'Great American Biotic Interchange' (Lessa
et al. 1997). The closure of the Isthmus of Panama and the development of a grassland corridor allowed interchange of large mammals.
The cyclical fluctuations of most taxa throughout the late Neogene
probably relate for the most part to the fluctuating sea-levels and
global climate, with parallels to the pattern seen in the late Neogene
of West Africa. In contrast Southeast Asia consists of a number of
tectonic plates which have moved and collided carrying biotas with
them and so a great deal of mixing of taxa from Asia and Australia
occurred.

2. A substantial proportion of the land area of Southeast Asia consists
of islands. Islands are long known to provide excellent opportunities
for speciation. The geological history of these islands, with repeated
isolations and connections to Australia and Asia, has also driven
speciation. In comparison to the fragmented pattern of islands that
one sees in Southeast Asia, Africa and South America have comparatively few islands and so allopatric speciation has been driven by
other forms of geographic separation.

Speciation

As well as the high level of endemism, another feature of Southeast
Asian fauna, exemplified by the avifauna and mammals, is the degree of
speciation and sub-speciation. For speciation to occur there usually has
to be some form of reproductive isolation. Following disconnection, the
process of evolution by natural selection (increased frequency of mutations
which enhance reproductive fitness) or genetic drift (a random change
in allele frequency resulting from the effects of chance on survival and reproductive ability of individuals in a small population) unfolds. Reproductively
isolated, the gene pools of the separated populations begin to diverge in genetic
composition in different ways. Speciation is more likely to occur when founder
populations are small in size, so genetic drift and the founder effect have more
consequences. The different selective pressures of the new environment
heighten the divergence caused by genetic drift.

Allopatric speciation occurs when a population becomes isolated geographically from the rest of the species. Although the evolution of new
species is almost entirely caused by allopatric speciation, sympatric speciation can occur if there is niche separation or temporal separation of

reproductive events, but no geographical separation. Geographic isolation can occur in several ways e.g. sea-levels rise, rivers change course, glaciation or formation of mountain ranges. In different groups of organisms, different geographic changes can create complete barriers, or none at all. For example, large rivers can be effective barriers for poor swimmers but not for amphibious taxa. The Cross River in eastern Nigeria is the eastern limit of the olive colobus monkey (*Procolobus verus*) and the western limit of several Congolese mammals, but is not an effective barrier for amphibians (Jenkins 1992). Brandon-Jones (1998) has also discussed the importance of poor swimming ability with regard to colonisation by rafting. An animal that is a poor swimmer, when on a raft of vegetation, will be more reluctant to leave the raft than a strong swimming species. The converse of this is that good swimmers are less likely to raft as they are more likely to abandon the raft before they are swept out to sea. Brandon-Jones (1998) uses this to explain both the endemism of the proboscis monkey (*Nasalis larvatus*) in Borneo and the absence of tigers (*Panthera tigris*) from Borneo, both able swimmers.

Southeast Asian forests and masting

Southeast Asian forests are dominated by mast-fruiting dipterocarps unlike any other major tropical forests. For example, over 50 species of Bornean dipterocarps employ the reproductive strategy termed masting (Curran *et al.* 1999). In much of tropical Southeast Asia, although there are some trees in flower and fruit all of the time, most species bear flowers and fruit only periodically, many of them annually, and some years are better than others. *Shorea macrophylla* and others produce a crop only once every few years. Many theories have been put forward to explain such masting or mass-flowering/fruiting events (Kelly 1994). The predator-satiation hypothesis seems easily the most tenable for tropical examples of masting. The contemporaneous flowering and fruiting of many canopy trees will result in seasonal peaks in food for herbivorous animals. As dormancy is short or zero, seed supply is sporadic. Secondly, seedlings will appear on the forest floor in occasional immense populations and so enhance their chances of maturation. Masting gives canopy trees an important survival advantage, because since so much seed is produced simultaneously over a large area, the predators are unable to consume all of the seed, leaving enough remaining to germinate and produce new seedlings. However, the irregular production of seeds may make it difficult for seed-eating mammals to maintain large populations in the forest. It can thus be advantageous for a species for individuals to flower synchronously. The strong synchrony of dipterocarp fruiting has major implications for forest ecology and conservation since these forests have many mobile seed predators (e.g. bearded pig, *Sus barbatus*) (Caldecott *et al.* 1993). In species that fruit gregariously and satiate their consumers, those individuals which reproduce late or early are more likely to have their fruits eaten. Thus natural selection will tend to promote and strengthen the simultaneous maturation of fruit (Turner 2001).

The evolutionary strategy for populations of long-lived trees is to maintain high genetic variability and one means to achieve this is by outbreeding i.e. cross-pollination between individuals. Successful pollination may also be enhanced by synchronous flowering, both in terms of numbers of pollinations and the likelihood of cross-pollination (Turner 2001). Synchronously flowering individuals of *Hybanthus prunifolius* showed higher fruit and seed set than those induced to flower at a different time from the main population (Augspurger 1981). Synchronous supra-annual flowering is a notable feature of the forests of west Malesia (Turner 2001). Over the course of a ten year study in Malaysia, the species which flowered in any given year varied every year, and included gregariously flowering Dipterocarpaceae. In fact, the flowering of Dipterocarpaceae is notorious for its infrequency and gregariousness. Records from Malayan Forest Department during 1925–70 reveal that dipterocarps, in general, fruit heavily every 2–3 years, with occasional intervals of up to five years, and with some differences between species. Even in the best years only 40–50% of mature trees in a given area are fertile and in some cases groups of trees flowered together. Many dipterocarps flower sporadically during any month of the year, and this has tended to obscure the existence of a single regular maximum flowering. Circumstantial evidence suggests that drought periods, 3–5 months before, promote mass flowering. Others suggest that it is not the drought itself that controls the phenology but the concomitant rapid increase in the hours of sunshine as wet, cloudy skies are replaced by dry weather with clear blue skies (Whitmore 1984). Intriguingly, dipterocarps in the peat swamp forest of Sarawak flower out of phase with those on dry land i.e. not in the dry season (Whitmore 1984).

According to Curran *et al.* (1999), the climatic conditions of an El Niño year trigger simultaneous fruiting in dipterocarps, and are essential for regional seed production. With El Niño events apparently occurring more frequently now than in the past, the masting trees are producing less mass of seed and the amount produced is not enough to satiate the seed-predators and so there is little production of new seedlings. The evidence suggests that the flowering of many lowland rain forest tree species is promoted by water stress, but the relationship is not simple. The amount of sunshine increases in dry periods so that could be more important than water stress because it induces a build up of assimilates (Whitmore 1984).

The uniqueness of the Southeast Asian biodiversity

The distribution of forest types within the Southeast Asian region is complex and is largely determined by three factors: (1) climate, (2) landform, and (3) human impact. Intergradations between different types can often occur even in relatively small areas resulting in mosaic-like patterns of vegetation cover.

The present climate of Southeast Asia is governed primarily by monsoonal winds (Bowman 2002). The northwest or winter monsoon picks up water from the west Pacific and South China Sea and is the cause of most of the rainfall in the Sunda region. The southeast or summer monsoon picks up water in the Indian Ocean bringing rain in the summer months. The oreographic (mountain-induced) rain-shadow effect in Sumatra, Java and Nusa Tenggara can make the summers comparatively dry (Whitmore 1984; An 2000).

The tropical closed canopy forests of the Asia–Pacific region are centred geographically in the Malesian groups of islands which lie between the Malaysian Peninsula and Australia. These islands were until relatively recently, covered mostly in rain forest, with a fringe of monsoon (seasonal) forests along the southern margin. To the north and west, rain forests extend up into continental Asia, where they occur in the wetter parts of Myanmar, Thailand, Cambodia, Laos and Vietnam, and form on the southern margins of China, Bangladesh, Assam and northeast India. There are detached fragments of rain forest in southwest India (the Western Ghats), in southwest Sri Lanka and the Andaman and Nicobar islands. At the northern and southern limits of the region, in areas of a seasonal climate, tropical rain forests gradually alter in floristic composition and become simpler in structure, becoming what is known as subtropical rain forests, and further still from the equator, they turn into temperate rain forests. These changes are accentuated by increases in elevation. Thus no sharp boundary can be drawn in upper Myanmar and in Assam between tropical rain forests in the lowlands and temperate rain forests in the high mountains of the southern Himalayas (Blower et al. 1991).

The Indo-Chinese floristic region has not been documented fully, but the total flora is estimated at c. 15 000 spp., at least one-third of which are endemic. Endemism is rather less at generic level. The Malesian flora is conservatively estimated at 25 000 spp. of flowering plants, about 10% of the world's total, with roughly 40% of the genera and even more of the species being endemic (McNeely *et al.* 1991). Myanmar has some tropical evergreen rain forests in the south, but most of its forests are semi-evergreen. There are varied floras with a strong Malesian influence in the far south. In the north there are temperate influences, and in the far north on the south slopes of the Himalayas the montane rain forest has a completely temperate flora with conifers and rhododendrons. About 7000 species of flowering plants have been recorded with 1071 endemics (Blower *et al.* 1991).

Anthropogenic effects

Humans and their human-like ancestors have lived in tropical Asia for at least 500 000 years, most of the time as hunter-gatherers – but even at this stage they had some impact on the environment (Rambo 1979). The alluvial flatlands, once covered by various kinds of freshwater swamp forest, have long ago been replaced by rice fields. In the seasonal monsoon forests of Indo-China civilisations have waxed and waned and some scientists believe that in these formations no significant pristine forest remains, all has been cleared for shifting agriculture at one time or another. There is a long history of shifting cultivation in Southeast Asia. Spencer (1966) came to the rather remarkable conclusion that virtually no pristine monsoon seasonal forest remained in Southeast Asia. Seasonal forest is more suited to shifting cultivation as a dry period makes it easier for burning. When shifting cultivation functions correctly, wildlife can flourish with many herbivores feeding in the abandoned areas and in turn predators are attracted by the herbivores. The older swiddens (areas of slash and burn agriculture) contain a high proportion of fruiting trees which attract squirrels, hornbills, primates and a wide range of other animals (Spencer 1966). Mature tropical forests conceal most of their edible products high in the canopy, beyond the reach of terrestrial herbivores. Wharton (1968) provided evidence that the distribution of large mammals in Southeast Asia, particularly the family Bovidae, was highly dependent on shifting cultivation. The seledang or gaur (*Bos gaurus*) prefers forest tracts adjacent to savannah or other human/fire affected open areas and is believed to have spread to Peninsular Malaysia by following shifting cultivation. Banteng (*Bos javanicus*) occupies secondary forest in humid areas and is confined to savannah woodland in deciduous forested areas. Banteng are also the probable ancestor of Asian domesticated cattle (Hedges 1996). The domesticated form, known as Bali cattle, are widely kept in Indonesia. Very little is known about the kouprey (*Bos sauveli*), but they are believed to show a preference for ancient, abandoned rice plantations, which undergo annual burning which maintains a savannah pasture (Wharton 1968). This relationship between the ecological requirements of large herbivores and human impacts on the landscape would seem to be an Asian phenomenon.

Comparison between the tropical biotas of Asia, Africa and South America

Approximately half of the world's area of tropical forests lie in the Neotropics, considerably more than in Asia (30%) or Africa (20%). In terms of species-richness for trees greater than 10 cm diameter at breast height (dbh), the Neotropics has the highest diversity, followed by Asia–Pacific and then Africa (Jenkins 1992; Harcourt et al. 1996; Morley 2000). The highest species diversities by unit area are also recorded from South America although very high diversities are also recorded from Indonesia and Malaysia (Morley 2000).

Africa is the driest of the three tropical continents (Collins 1992). It has long been known that African tropical forests contain fewer animal and plant species than Southeast Asia or South America (Haffer 1974; Hamilton 1982). It has been suggested that there was once similar diversity in Africa but the effects of Quaternary aridity were much greater there and so a greater number of extinctions occurred. This climate change was more pronounced in Africa due to topography. There were fewer and more fragmented moist forest refugia which were relatively small in area and so suffered elevated extinction rates. There were several forest refugia in Africa, small areas in Sierra Leone/Liberia and East Ivory Coast/West Ghana and larger refugia in Cameroon/Gabon, East Zaire and East Tanzania (Jenkins 1992). The endemic plant genera of the moist tropical African forests contain fewer species in contrast to endemic genera elsewhere in the tropics. There were a greater number of refugia in South America and in Southeast Asia and for Southeast Asia, the large proportion of islands, with the proximity of the sea to many forests had a buffering effect.

Like the flora, the Neotropics are extremely rich in wildlife. South America has the highest number of land birds; about 3300 species. For mammals, there are similar overall numbers between Africa and South America in terms of numbers of species, genera and families, especially in rodents and primates. A substantial area of tropical Africa is savannah with a large number of species and vast numbers of large herbivores i.e. artio-dactyls (mainly antelope) and perissodactyls (Bourliére 1973; Long et al. 1996; Redford et al. 1996). There is nothing comparable in Asia and the South American grasslands are very species poor for large mammals. Carnivores are present in all three regions with Panthera genus cats found in all. There are great similarities between African and Asian faunas (e.g.

apes, rhinoceros, elephants, pangolins, hornbills, babblers, monitors, pythons). There are even some species in common. Two disparate examples are the leopard and the ring-necked parrakeet (*Psittacula krameri*), although both exist as separate sub-species. However, there are also similarities between Asia and South America, with tapirs common to both, but Procyonoid carnivores are only found in the Americas with the exception of the red panda (*Ailurus fulgens*) of Nepal, Tibet, China, Bhutan and northern Myanmar. Bears (Ursidae) are widespread in Southeast Asia, Eurasia, North America and Andean South America but are absent from Africa. The wrens (Troglodytidae) are an American family with one species (*Troglodytes troglodyte*) widespread in the Holarctic but absent from Africa. Marsupials are widespread in the Neotropics, Australasia and just into Southeast Asia (Sulawesi, New Guinea and Malaku).

The current state of biodiversity and conservation

The dipterocarp forests of Southeast Asia are the largest source of timber for the world's trade. Although the area of forest destroyed each year in the Americas is much higher than elsewhere, the annual percentage cleared is highest in Asia (Harcourt *et al.* 1996; see Chapter 1). The International Union for the Conservation of Nature and Natural Resources have designated three plant and eight animal species as 'extinct' in Southeast Asia (IUCN 2003). The history of large-scale deforestation in Southeast Asia can be described as recent (i.e. over the past two centuries). Many of the native species of the region, although presently extant, might be persisting as 'living dead', doomed to extinction (such as rare long-lived trees), the fragmentation of their habitats leading to reproductive isolation (Brook *et al.* 2003). So although the actual number of extinct species in the region is not currently alarming, the level of threat (e.g. by deforestation) to extant species now imperils the biota on a regional scale (detailed in Chapters 3, 4 and 5). The number of threatened species in Southeast Asia, including those in the IUCN categories of 'critically endangered' (CE), 'endangered' (EN) and 'vulnerable' (VU) ranges from 20 (CE) to 686 (VU) species for vascular plants, 6 to 91 species for fish, 0 to 23 species for amphibians, 4 to 28 species for reptiles, 7 to 116 species for birds, and 5 to 147 species for mammals (IUCN 2003). The loss of many of these regional populations would be likely to result in global extinctions because of the very high proportion of endemic species (Brook *et al.* 2003; WRI 2003). For example, of the 29 375 vascular plant species in Indonesia, almost 60% do not occur anywhere else (WRI 2003; Sodhi *et al.* 2004b).

The last few years has seen a number of species new to science being discovered, especially in the relatively poorly studied Laos–Vietnam–Cambodia region, and include the golden-winged laughingthrush (*Garrulax ngoclinhensis*) (Eames 1999a) and the black-winged barwing (*Actinodura sodangor*) both from Vietnam (Eames 1999b). There have also been rediscoveries of wild specimens of species once thought to be extinct in the wild, such as two pheasants, Edward's pheasant (*Lophura edwardsi*) in 1996 and the imperial pheasant (*L. imperialis*) in 1990 and 2000 (Madge & McGowan 2002).

There have also been recent discoveries of new species of large mammals from the same region. The Annamite Mountains of central Laos and Vietnam have yielded: the saola (*Pseudoyrx nghetinhensis*), discovered in 1992 and described in 1993 (its relationship to other bovids is unclear); and three new deer species, the large-antlered muntjac (*Muntiacus vuquangensis*) in 1994, Annamite muntjac (*M. truongsonensis*) in 1998, and Putao muntjac, (*M. putaoensi*) in 1999 (Francis 2001). A new species of civet *Viverra tainguensis* has been described from specimens collected in the 1980s from Vietnam (Kanchanasakha *et al.* 1998).

Given that the taxonomic groups of birds and mammals are comparatively well known and usually highly conspicuous, and yet new species (to science) are still being discovered, the obvious implication is that for less well studied taxa there is undoubtedly much more to learn and many species awaiting description. For example, in Laos almost 300 new species of fish have been described since 1996 (Kottelat 2001). In Southeast Asian mainland hills, 1000 species of freshwater fish are known, of which 500 are endemic. It is estimated another 200–500 species are still unknown. In Malayan, Bornean and Sumatran coastal peat swamps and swamp forests (habitats of restricted area, largely unstudied and traditionally thought to be species poor), 100 endemic fish species have been described. From the Lower Mekong 121 species of gastropod molluscs are known with 111 being endemic. From Myanmar-Malayasia 100 species of freshwater crabs in 30 genera in three families are so far described (United Nations Environmental Programme undated). Many of these species have very restricted ranges and habitats and so extinction, perhaps before they are ever described scientifically, is a very real threat.

Summary

Southeast Asia has an extraordinarily rich biodiversity and endemism across a wide range of plant and animal taxa for the following reasons:

1. The geological history involved the collision of tectonic plates from Australasia with Asia, bringing and mixing their respective faunal and floral elements.

2. During periods of intense glaciation (ice-ages), the lowering of sea-levels exposed a massive continental shelf that reached from northern Australia to Taiwan, facilitating migration. Sea-levels rose and fell a number of times thus driving speciation, with the final rise event ending around 7000 years ago.

3. The geography is important, firstly, in being largely part of the huge Eurasian land mass with pathways for dispersal from north, central and west Asia, Europe, and Africa via the Middle-East.

4. The second geographical aspect is the present-day highly insular nature of much of the region. This is in contrast to South America and Africa that have comparatively few islands. Islands, by their very nature isolate biotas and thus promote speciation and diversification.

5. The proximity of many of the regions' tropical forests to the sea meant that during periods of lower global temperatures there was a maritime buffering effect on temperatures and rainfall, permitting greater survival of refugia.

6. Anthropogenic factors that influence the landscape are of differing historical types and impacts in Asia compared to those in Africa and South America. For instance, although both of the latter have had recent catastrophic deforestation, this has only been in the last 200 years. Southeast Asia has had a much longer period of forest change with time for some species e.g. bovids, to adapt and become dependent on human-influenced landscapes.

7. Like all of the world's tropical forests, those in Southeast Asia are highly threatened by contemporary human activities. The high degree of island endemism in this region makes the probable future for many species particularly bleak.

Chapter 3

Biotic losses and other effects
of habitat degradation

Using terrestrial biotas, we now move to the task of illustrating the adverse impacts of habitat loss and degradation on Southeast Asian biodiversity. One key point to bear in mind is that the biodiversity of Southeast Asia has been relatively poorly studied compared to that of other tropical or sub-tropical regions (e.g. Central America and Caribbean, South America, and Sub-Saharan Africa), particularly over the past 20 years (Sodhi & Liow 2000; Sodhi *et al.* 2004b) (refer to Fig. 1.11). In particular, some taxonomic groups and habitats within Southeast Asia remain especially poorly studied. For example, there is an alarming dearth of information on vascular plants (Fig. 3.1), even though they constitute the vast majority of the biomass, as well as habitat types such as caves (Fig. 3.2). Also vertebrate populations are apparently declining faster than their habitat (Balmford *et al.* 2003a) which suggests that the documented biological impacts of habitat disturbance may be grossly underestimated.

Here we present the effects of habitat degradation on biotic groups such as plants, arthropods, amphibians, reptiles, birds and mammals, using numerous Southeast Asian examples. We also discuss briefly the extinction-proneness of the biota and effects of habitat degradation on ecological interactions and specialised behaviours.

Impacts on biotas
Plants
Plants are obviously the first organisms to be impacted by deforestation and forest disturbance. For instance, since 1819, Singapore has lost 99% of its primary forest and the original mangroves (Turner *et al.* 1994; Hilton & Manning 1995), which led to a consequent loss of 26% of its original 2277 native vascular plant species (Turner *et al.* 1994; Brook *et al.* 2003). Because Singapore is a land-bridge island with low endemism, these losses represent local and not global extinctions, but the habitat loss–species loss relationship

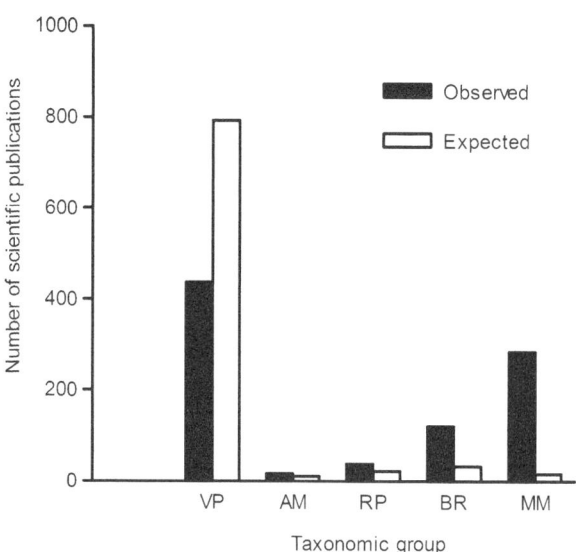

Figure 3.1 Relative biodiversity research effort among six taxonomic groups in Southeast Asia, including vascular plants (VP), amphibians (AM), reptiles (RP), birds (BR) and mammals (MM). See Fig. 1.11 to see how expected values were calculated.

is still disturbing. Coastal habitats (including mangroves) and inland forests lost 39% and 29% of their vascular plant species, respectively (Turner *et al.* 1997). Some 88% of the original 196 native orchid species vanished. Similarly, 62% of the original 297 epiphyte species have been lost. High losses in orchids and epiphytes may be due to factors such as the loss of preferred hosts (big trees), microclimate changes brought about by fragmentation and/or over-exploitation for the horticultural trade (Turner *et al.* 1994).

Considering that some plants may live for centuries, many of the currently extant species in Singapore may nevertheless be committed to extinction due to unviable populations (i.e. living dead) (Turner *et al.* 1994). This has been illustrated by a study of plant extinctions from an isolated 4 ha fragment of lowland rain forest in Singapore (Singapore Botanic Gardens) (Turner *et al.* 1996). Although 49% of 448 vascular plant species recorded in the 1890s were extirpated by 1994, more shorter-lived shrubs (74%), climbers (61%) and epiphytes (67%) were lost than long-living trees (42%) (Fig. 3.3). However, half of the tree species were represented by only one or two individuals (>5 cm dbh) indicating that their populations are doomed.

In addition to forest loss, forest disturbance, such as logging for timber, can negatively impact tree species-richness and tree recruitment (Slik *et al.* 2002; Foody & Cutler 2003). But do plant communities recover after

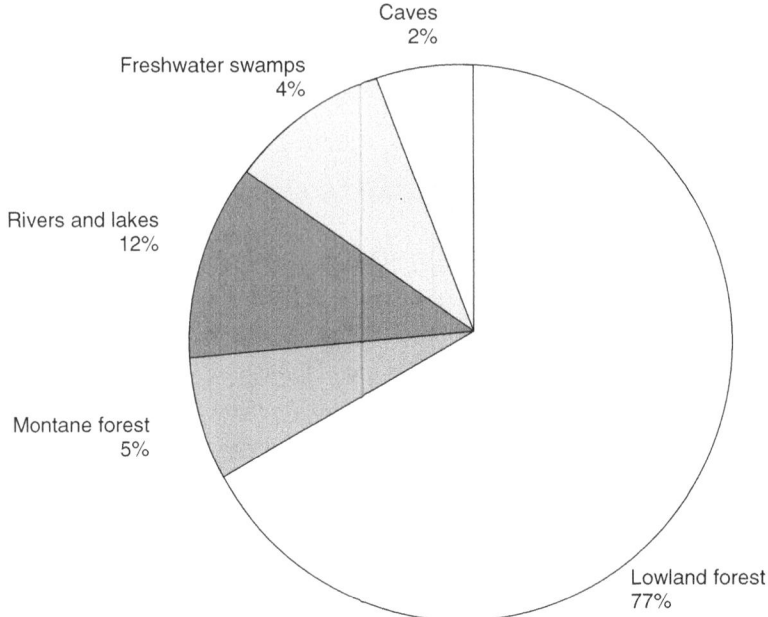

Figure 3.2 Relative biodiversity research effort among different habitats in Southeast Asia. We searched internationally peer reviewed scientific publications from the database *Biological Abstracts* (*BIOSIS Previews*), which contains over 13 million records from 1969 to 2004. These citations have been gathered from over 5500 sources, featuring traditional life sciences (e.g. biology, botany, zoology), interdisciplinary (e.g. agriculture) and other related topics (e.g. methods).

disturbance? To answer this question, Turner *et al.* (1997) sampled 0.2 ha plots in primary and 100-year-old secondary forests in Singapore. They found that 16 plots in the primary forest contained a total of 340 species of trees (>30 cm dbh) whereas 43 plots in the secondary forest contained only 280 species. Further, the mean species-richness in secondary forests reached only 60% of that in the primary forest, even after a full century of succession. It is possible that either environmental conditions (e.g. soil nutrient levels) were not conducive for the growth of late successional primary forest trees in the secondary forest, or that their seed simply did not arrive there. Seeds of the secondary forest species are primarily dispersed by common birds of the open country (e.g. yellow-vented bulbul, *Pycnonotus goiavier*), whereas relatively large seeds (>70 cm long) of characteristic primary forest trees require larger frugivores (e.g. hornbills) for seed dispersal (Corlett 1991; Corlett 1992). Such frugivores have already been extirpated or are very rare in Singapore (Castelletta *et al.* 2000),

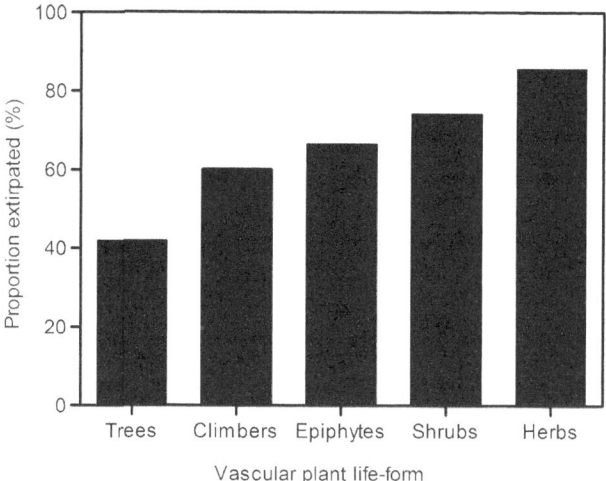

Figure 3.3 Proportion of different vascular plant life-forms that were extirpated from the Singapore Botanical Garden. (Data from Turner *et al.* 1996.)

highlighting one of the serious knock-on effects of biodiversity loss – a diminution in the ability of degraded or cleared habitat to recover, even in the long-term. Another agent of seed dispersal, that is wind, may not be able to disperse the primary forest dipterocarp seeds far from the primary forest. This may be one of the reasons why wind-dispersed primary forest plants have not spread extensively in heavily urbanised Singapore (Turner *et al.* 1997).

In circumstances somewhat similar to what has been observed in Singapore, in the rain forests of Peninsular Malaysia (Pasoh Forest Reserve), the mean canopy height (24.8 versus 27.4 m), mean canopy surface area (12.0 versus 17.4 m^2), mean crown size (42.9 versus 94.5 m^2) and diversity of trees (Fisher's α: 110.1 versus 122.1) were significantly lower in a logged forest (41 years old) than a neighbouring primary forest (Okuda *et al.* 2003). These results show that tree diversity, canopy height and volume in a regenerating forest matches poorly with those in the primary forest, even after more than four decades of recovery.

Is selective logging detrimental to tropical plant communities? To address this question, a study was carried out on the effects of selective logging (removal of approximately one-third of all trees) on vegetation in Kalimantan (Gunung Palung National Park; Indonesian Borneo) (Cannon *et al.* 1994). They found that harvesting removed 62% of dipterocarp basal area. Dipterocarp trees less than 50 cm in diameter suffered high mortality due to logging operations, and this family may have limited future commercial potential. A later study from the same area showed that there was an increase in tree diversity eight years after selective logging (Cannon et al. 1998). Although there were fewer species of trees more than 20 cm in diameter in eight-year logged plots than in similar-sized unlogged plots, their mere presence indicates that these could be reproducing individuals thus contributing to regeneration. It has been argued, however, that these results should be viewed with caution, because different species have different inherent conservation values (Sheil et al. 1999). From a conservation context, we should be more concerned whether eight-year selectively logged sites can retain or recruit disturbance-sensitive taxa that characterise old-growth forest.

Removal of some types of trees during logging operations can also impact on animals that rely on these. Logging generally reduces the availability of nesting trees (e.g. Vitex spp.) for hole-nesting vertebrates (e.g. woodpeckers) (Pattanavibool & Edge 1996). Large cavity forming tree species need to be spared during tree harvesting to cater for the needs of hole-nesting vertebrates.

Skid trails created during selective logging can remain devoid of trees. By taking proper precautions during logging operations (that is minimising skidding and unnecessary tree felling), establishment of pioneer seedlings can be enhanced (Howlett & Davidson 2003). In areas where skid trails and log landings have already been created, forest regeneration can be facilitated by adding seeds of pioneer trees (Pinard et al. 1996). The chances of success of such management options can be improved by appropriate site preparation (e.g. by creating adequate but not too large gaps in the canopy) (see Chapter 6) to enhance the survival of pioneer tree seedlings. Fertilisation by nutrients such as nitrogen and phosphorus could also be used to enhance tree growth, although nutrient addition experiments have not so far shown significant girth increments in either dipterocarp or non-dipterocarp trees in Kalimantan (Barito Ulu) (Mirmanto et al. 1999).

Crown liberation (cutting and girdling of pioneer species) is a forest management practice applied in certain areas such as Kalimantan. This is done following logging to enhance the growth of primary forest tree seedlings and to facilitate succession of the forest towards its pre-logging state. However, it has been found that such a practice may be counterproductive, as it only favours recruitment of the seedlings of light-demanding pioneer species rather than the intended target – primary forests' dipterocarp trees (Kuusipalo et al. 1996). Instead, Kuusipalo et al. (1996) recommended that crown liberation, removing shading trees over areas with abundant dipterocarp seeding may be more cost-effective. Such artificially created vegetation gaps (400–600 m^2) can thus assist in natural regeneration. In a later study from the same area (Kintap, Kalimantan), it was found that such gaps had a total dipterocarp basal area of 57% more than control, untreated areas six years after logging (Kuusipalo *et al.* 1997). Further, survival of dipterocarps was 33% higher and their diameter twice as high in the artificially created gaps than in untreated control plots. These results strengthen their recommendation of gap creation as a feasible forest management practice. If artificial gaps are created to facilitate regeneration, it must nevertheless be kept in mind that success by different dipterocarp species can be dependent upon interspecific competition (Brown *et al.* 1999).

Because degraded areas can be rich in potential predators (Sodhi *et al.* 2003), seed predation can affect forest regeneration in these areas. An earlier study reported relatively high (98%) predation of artificial seeds (peanuts, *Arachis glabrata*) in primary and secondary forest patches of Singapore (Wong *et al.* 1998). Similar results (72% predation) were found when jackfruit (*Artocarpus heterophyllus*) seeds were placed in Singaporean forest patches (Sodhi *et al.* 2003). In a similar result from Linggoasri, Java, predation of artificially placed jackfruit seeds was at least 20% lower in selectively logged forest than in regenerating forests (3 years old) and exotic pine (*Pinus merkusii*) plantation (Sodhi *et al.* 2003). It is unclear if artificial seed predation studies accurately mimic predation of natural seeds, but if they do, regeneration may be seriously impacted due to high seed predation in disturbed forests.

Other studies also support these reported trends of high seed predation in disturbed forests. The diversity of seedlings and saplings in the rain forests of Sabah (Danum Valley Field Centre, Malaysia) was shown to be associated negatively with the intensity of logging activities (Foody & Cutler 2003). Artificially planted *Shorea stenoptera* seeds experienced higher predation in logged than in primary forest in Kalimantan (near Gunung Palung National Park (Curran & Webb 2000). This result indicates high seed predation in at least some logged areas. Exclusion of mammalian herbivores enhanced survival (by 38%) and growth (44% taller) of pioneer seeds in the same site in Sabah (Howlett & Davidson 2003). Therefore, predator exclusion can be a viable management option for promoting the regeneration of forests.

Arthropods

Many studies have been conducted to document the effects of habitat loss and degradation on arthropod faunas, the most diverse and abundant of all macroscopic taxa. However, the greater proportion of these studies from Southeast Asia has been from just one site: Danum Valley Field Centre

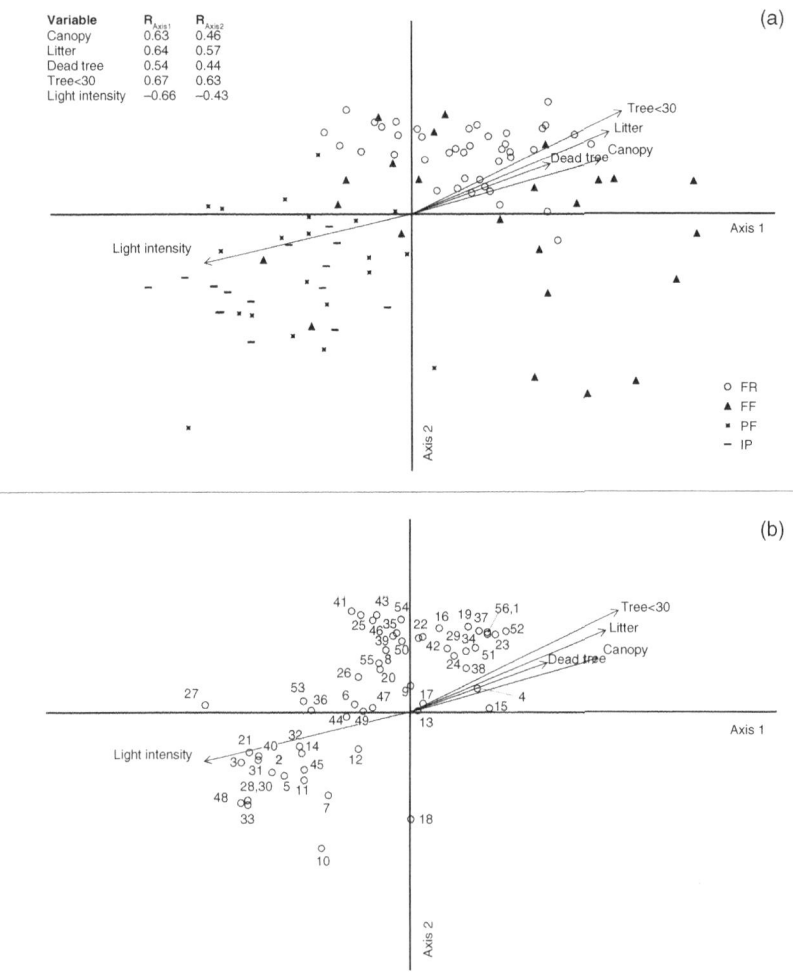

Figure 3.4 NMS ordination joint plot of sample scores (a) and species scores (b) with the significantly correlated ($R > 0.50$) environmental variables, number of trees less than 30 cm dbh (Tree < 30); mean leaf litter depth (Litter); number of dead trees (Dead tree); canopy cover (Canopy); and light intensity (Light intensity). FR, FF, PF and IP represent forest reserves, forest fragments, urban parks adjoining forests and isolated urban parks, respectively. 1 = *Amathusia phidippus*; 2 = *Appias libythea*; 3 = *Catopsilia pomona*; 4 = *Catopsilia pyranthe*; 5 = *Catopsilia scylla*; 6 = *Cethosia hypsia*; 7 = *Chilasa clytia*; 8 = *Cirrochroa orissa*; 9 = *Cupha erymanthis*; 10 = *Danaus chrysippus*; 11 = *Danaus melanippus*; 12 = *Delias hyparete*; 13 = *Doleschallia bisaltide*; 14 = *Elyminas hypermnestra*; 15 = *Elymnias panthera*; 16 = *Eulaceura osteria*; 17 = *Euploea mulciber*; 18 = *Euploea phaenareta*;

(Sabah). While these studies are important in understanding the congruence among taxonomic groups in their response to disturbance, studies from other sites are needed urgently to test whether assumed 'generalities' can validly be drawn and extrapolated from the studies centered around this site.

Butterflies are commonly used as an invertebrate indicator taxon for ecological research. They are easy to observe and have been found to be sensitive to habitat disturbance, thus making them a suitable ecological 'miner's canary'. In Singapore, due primarily to deforestation, 61% of the original 381 butterfly species have been extirpated since 1819 (Brook *et al.* 2003). The island's existing forest reserves (protected old secondary and primary forests) retain a higher total number of species, number of unique species, density of species and community evenness than do the scattered rural vegetation fragments and urban parks (artificially re-vegetated habitats) (Koh & Sodhi 2004). Forest reserves contained 72% of 58 species recorded by Koh & Sodhi (2004). This shows that the continued preservation of the reserves is necessary to maintain the extant forest butterfly species in Singapore. Further, different butterfly species in Singapore respond to different environmental variables (e.g. canopy closure) suggesting that environmental heterogeneity may be needed for the conservation of large and diverse numbers of butterfly species (Fig. 3.4).

As might be expected, forest dependent butterfly species (e.g. *Elymnias panthera*) generally avoid urban parks and forest fragments (Koh & Sodhi 2004). This study recommended that urban parks can be vegetated with suitable butterfly host plants (e.g. *Uncaria gambir*) to provide a more attractive habitat for the butterflies.

There are a number of studies highlighting the vulnerability of restricted range butterfly species to habitat degradation. In the dry forest of Thailand

Figure 3.4 (*cont.*)
19 = *Euploea radamanthus*; 20 = *Eurema hecabe*; 21 = *Euthalia aconthea*;
22 = *Euthalia monina*; 23 = *Faunis canens*; 24 = *Graphium agamemnon*;
25 = *Graphium evemon*; 26 = *Graphium sarpedon*; 27 = *Hypolimnas anomala*;
28 = *Hypolimnas bolina*; 29 = *Idea stolli*; 30 = *Ideopsis vulgaris*; 31 = *Junonia almanac*; 32 = *Junonia hedonia*; 33 = *Junonia orithya*; 34 = *Lasippa tiga*;
35 = *Lebadea martha*; 36 = *Leptosia nina*; 37 = *Lexias pardalis*; 38 = *Moduza procris*; 39 = *Mycalesis sp.*; 40 = *Papilio demoleus*; 41 = *Papilio demolion*;
42 = *Papilio iswara*; 43 = *Papilio memnon*; 44 = *Papilio polytes*;
45 = *Parantica agleoides*; 46 = *Pathysa antiphates*; 47 = *Phaedyma columella*;
48 = *Phalanta phalantha*; 49 = *Polyura hebe*; 50 = *Tanaecia iapis*;
51 = *Tanaecia pelea*; 52 = *Thaumantis klugius*; 53 = *Troides helena*;
54 = *Vindula dejone*; 55 = *Ypthima sp.*; 56 = *Zeuxidia amethystus*. (Reprinted with permission from Koh & Sodhi 2004.)

(Huay Kha Khaeng Wildlife Sanctuary), it was found that logging has impacted negatively upon the area's butterfly fauna (Ghazoul 2002). Although similar numbers of butterfly species (37–40) were found in unlogged and nearby logged areas, the highest butterfly abundance was at the former. Species with restricted geographic distribution occurred more frequently in unlogged sites, suggesting that logging may be particularly detrimental to these range constrained species.

Other studies also report the vulnerability of restricted range butterfly species. A comparison of the butterfly fauna of unlogged and logged (5 years prior to the study) habitats was carried out on Buru Island (Indonesia) (Hill *et al.* 1995). Of the 41 species recorded, 25 were common to both areas. A total of 12 species were only found in unlogged forest, including two endemic species (*Delis rothschildi* and *Mynes talboti*). Species were abundant and more evenly distributed in unlogged rather than logged forest. Therefore, selective logging seemed to affect, negatively, the endemic forest butterfly fauna of this island. Similarly, in central Sulawesi (Lore Lindu National Park, Indonesia), richness and abundance of butterfly species endemic to the Sulawesi region were higher in undisturbed than in disturbed forest (including farm and regenerating forest) (Fermon *et al.* 2003).

Butterfly fauna of four sites in Sumba (Indonesia) were compared (Hamer *et al.* 1997). These sites differed in the level of protection and associated level of human disturbance. Butterfly diversity was highest in the most disturbed secondary forest. However, species with restricted ranges attained highest densities in the undisturbed primary forest. This study, and others like it (e.g. Hill *et al.* 1995; Ghazoul 2002) highlight the message that undisturbed primary forests are necessary if we are to conserve range restricted butterfly species. Such species certainly have high value in terms of maintaining global biodiversity, and perform important ecological services such as pollination of forest trees.

Does modest disturbance such as selective logging always harm butterflies? In Sabah (Danum Valley Field Centre), research was undertaken to determine the effects of selective logging on butterfly fauna (Willott *et al.* 2000). More butterfly species (151 versus 121) and individuals (1723 versus 1543) were found in the logged rather than unlogged forest. There seemed to be no taxonomic bias in the distribution of the butterflies: the number of species within families was similar in the logged and unlogged areas. Unlike Hill *et al.* (1995) and Ghazoul (2002), this study found that a low-level of selective logging (total removal of 3–7% of trees, with some damage to remaining trees) had little negative impact on the butterfly fauna of this site. Perhaps the close proximity to the unlogged forest (a likely source area) assisted in maintaining high butterfly richness in the logged forest.

Relatively little information is available on how individual species respond to disturbance. A comparison was made of the spatial distribution, abundance and habitat requirements of *Ragadia makuta* in neighbouring unlogged and logged (8–9 years ago, 7% of the trees removed for commercial purposes but 60–80% of them were destroyed due to logging operations) forests in Sabah (Danum Valley Field Centre) (Hill 1999). This species had a track record of being a good indicator of disturbance, being found to associate strongly with closed-canopy forest. Hill (1999) found that this butterfly species was restricted to less dense forest close to streams in her study area, although this could be due to the fact that the study coincided with a severe drought. Spatial distribution and abundance of this butterfly did not differ between logged and unlogged habitats. This was reflected by the fact that suitable habitats for this butterfly were still available in the logged areas. This study suggests that with the availability of suitable habitats, selectively logged forest can fulfill the ecological requirements of some sensitive wildlife species. A later study found that for *Ragadia makuta*, both drought and heavy rainfall caused population declines (Hill *et al.* 2003). Populations did recover after the weather became more favourable, and there were no differences in the population recovery between logged and unlogged areas, implying that selective logging may not be overly damaging for this species.

Forest gaps are created by natural tree fall or through forest management (see above). Are these gaps of any conservation value? To address this question, traps (baited with rotting fruits) were placed in forest gaps (canopy openness of 16%) and in closed-canopy (canopy openness <1%) areas of a forest in Sabah (Danum Valley Field Centre) (Hill *et al.* 2001). Although similar numbers of species were found in both sites (34 and 36), closed-canopy sites had species with more restricted geographic distributions. Species in the gaps had relatively broader and larger thoraxes, indicating that they enjoyed superior flying abilities. These results show that forest gaps contained more widespread and mobile butterfly species than closed-canopy areas. Therefore, canopy opening might be more harmful to the forest butterfly species with restricted distributions. In a subsequent study from the same area, it was reported that species responses to gaps and selective logging may be taxonomically biased (Hamer *et al.* 2003). Species of the families Satyrinae and Morphinae which had higher shade preference and narrower geographical distributions were affected adversely by selective logging, whereas the high light preferring cosmopolitan species benefited. On the other hand, in Nymphalinae and Charaxinae, widely distributed species were negatively impacted by logging but species with restricted geography benefited. Hamer *et al.* (2003) argued that these trends were observed because of loss of habitat heterogeneity through logging (e.g. lack of areas with dense shade). A mosaic of dense shade and open gaps in unlogged forest attracted species with opposing habitat requirements. Therefore, the conservation value of logged areas can be enhanced by increasing habitat heterogeneity. However, the type of disturbance (e.g. logging versus burning) may impact butterfly richness differently, and this fact should be kept in mind while evaluating the conservation value of areas for butterflies (Cleary 2003).

Do moths respond to habitat disturbance in a similar fashion to butterflies? It has been suggested that forest moth fauna can recover 100 years after disturbance if the area is not degraded further (Holloway & Barlow 1992). The effects of selective logging on moth fauna were studied in Sabah (Danum Valley Field Centre) (Willott 1999). Using light traps, Willott (1999) sampled primary forest understorey, canopy and selectively logged forest. The number of moth species varied between 352–458, 390–391 and 368–393 in the primary forest understorey, canopy and selectively logged forest, respectively. This result shows that there was no evidence for the hypothesis that the canopy of the primary forest contained the highest species-richness or that the moth fauna was depauperate in the selectively logged forest. However, after accounting for species' abundances, Willott (1999) predicted that 10% of the moth species confined to the primary forest may be lost due to logging in the study area.

Other studies also show that moths have some hope of maintaining viable populations in disturbed areas. Primary forests had higher (Fisher's $\alpha = 75$–128) geometrid moth diversity than agricultural areas and most secondary forests (34–61) in Sabah (Mt. Kinabalu National Park) (Beck et al. 2002). However, another 15 year old secondary forest had geometrid moth diversity comparable to primary forests (89), because of its well developed undergrowth. Corroborating this trend, undergrowth plant diversity proved to be the best predictor of geometrid diversity, suggesting that such habitat should be preserved if a high species richness of moth fauna is to be maintained. This conclusion is perhaps not altogether surprising, as moths are herbivores and in addition need plants on which to lay their eggs. In another study, the geometrid moth diversity in Peninsular Malaysia (Greater Templer Park Area) was the highest in a secondary forest that was selectively logged than in regenerating forest that was clear cut and abandoned around a tin mining area (Intachat et al. 1997). This study shows that secondary forests have a conservation potential for moths, but it is not clear how good they are relative to nearby large primary forests.

How does forest degradation affect other arthropod groups? As major decomposers, termites play a vital role in maintaining soil stability,

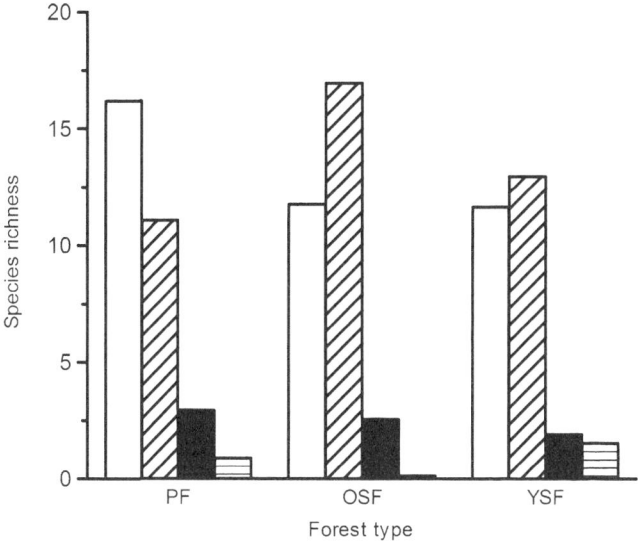

Figure 3.5 Species-richness of feeding groups in each of the three forest types, primary forest (PF), old secondary forest (OSF) and young secondary forest (YSF). Totals are means of two transects per forest type. Soil-feeders = unhatched bar; wood-feeders = diagonal hatched; soil-wood feeders = black bar; litter-feeders = horizontal hatch. Lichen-feeders were found only in YSF. (From Eggleton *et al.* 1997. Reproduced by courtesy of GTÖ – Society for Tropical Ecology.)

structure and quality in tropical forests. Soil-feeding termites specialise on soil-organic matter that usually gets degraded by disturbance. Further, soil feeding termites may have lower dispersal abilities than other termite groups (Davies *et al.* 2003). The effects of logging on termite fauna was determined in Sabah (Danum Valley Field Centre) (Eggleton *et al* 1999). Primary forest and two selectively logged forests (17 years and 3 years prior to the study) were sampled. Depending upon the type of sampling method used, primary forest had 3–5 or 1–2 more species than the 3 and 17 year old logged forests, respectively. Soil-feeding termite species-richness was highest in the primary forest whereas that of wood-feeding termites was highest in 17 year old logged forest (Eggleton *et al.* 1997) (Fig. 3.5).

Other studies on termites also document similar results. An assessment of the composition of termite assemblages across a disturbance gradient (agricultural land to primary forest) was undertaken in Sumatra (Jambi Province, Indonesia) (Jones *et al.* 2003). Termite species-richness and relative abundance were the highest in the primary forest and lowest in the cultivated area. Out of 34 species found in the primary forest, only one was found in the cultivated area. As with the study from Cameroon (Africa) and Sabah (Eggleton *et al.* 1995; Eggleton *et al.* 1999), soil-feeding termite species showed an impoverishment in their diversity with increasing disturbance. Jones *et al.* (2003) also found that woody plant basal area was correlated strongly with termite species-richness and relative abundance. They hypothesised that reduction in canopy cover and associated effects (such as alteration of microclimate and loss of feeding and nesting sites) may have caused the impoverishment of termite fauna in disturbed areas. To mitigate the loss of termites, Jones *et al.* (2003) recommended the implementation of actions such as low-impact logging (e.g. avoiding damage to non-targeted trees during logging operations), leaving dead wood to decay in logged areas, and protection of large forest patches.

Termite diversity was higher in primary forest (MacArthur's diversity index = 2.896) than 40 year old regenerating forest (2.036) in Peninsular Malaysia (Pasoh Forest Reserve) (Takamura 2003). This trend may be due to the lack of suitable (larger) foraging and nesting trees in the regenerating forest. A review using data from various studies, on the effects of human disturbance on Southeast Asian termites was undertaken (Gathorne-Hardy *et al.* 2002a). It was found that termites were affected strongly by disturbance, particularly by the total clearance of the forest canopy. If a disturbed area close to a primary forest is left untouched for approximately 50 years, it can recover completely its lost termite fauna. This is welcome news, because termites can be critical to the recovery of soil fertility following habitat disturbance. Gathorne-Hardy *et al.* (2002a) did, however, warn that termite faunal diversity and their vital ecosystem functions may be altered severely if current widespread rain forest loss in Southeast Asia is not abated.

In addition to maintaining soil health, termites may release large quantities of methane, carbon dioxide and molecular hydrogen. Methane is an important 'greenhouse gas' with a potential to impact earth's radiation balance. There is a concern that an increase in termite densities in anthropogenically modified areas may cause an elevation in the amount of termite-generated methane released into the atmosphere (Zimmerman *et al.* 1982). Eggleton *et al.* (1999) determined the gas physiology and landscape gas fluxes of termites, and concluded that it is unlikely methane production by termites in Southeast Asia contributes significantly to global increases in methane production (MacDonald *et al.* 1999). This deduction requires further verification, however, taking into consideration that habitat disturbance can alter species composition and that methane production can vary across different species (Jeeva *et al.* 1999).

Like termites, bees are critical in ecosystem functioning, being the dominant pollen vectors in tropical forests. With the rapid decline of forests in

Southeast Asia, bees become increasingly critical for forest regeneration and restoration. Bee communities were studied in forests along a disturbance gradient (primary forest through to plantations) in Peninsular Malaysia and Singapore (Liow *et al.* 2001). Below-canopy bee abundance, particularly of the family Apidae, was found to be significantly higher in primary forests than other types of habitats (Fig. 3.6). The distribution of bees was influenced by variables such as the density of large trees (diameter 30–40 cm), ambient temperature and flowering intensity of trees and shrubs. These variables are indicative of forest disturbance and resource abundance.

Similar patterns are reported by other workers. Bees were sampled from various localities in Sumatra (Indonesia) (Sakagami *et al.* 1990). They found that primary forests had higher bee richness, diversity and abundance than disturbed areas (e.g. plantations). More bumble bee species were found in a primary forest than secondary and disturbed forests. As with some termite species, availability of nest sites may be a limiting factor in bee distribution, with those species nesting in the cavities of large trees (e.g. *Trigona canifrons*) being restricted to the primary forests. The importance of suitable nest sites for bees was also highlighted in a study from 14 sites in Sabah (Eltz *et al.* 2002). They found that the nest density of stingless bees was lower in the primary forest than in forest fragments or plantations. They suggested that abundance of these bees was primarily dependent upon local availability of food resources (pollen). Mean number of pollen was surprisingly not the highest in the primary forest, and some of the species (e.g. *Trigona collina*) appeared to rely on non-forest pollen (e.g. from crops and mangroves), especially when little flowering takes place in the forest. However, 86% of the nest trees were large (>60 cm dbh) and about 40% of these were potential harvest trees (Eltz *et al.* 2003). Similar results have been found from Sarawak (Upper Baram Area) (Samejima *et al.* 2004): the nesting density of stingless bees was positively correlated with the density of large trees (>50 cm dbh). Therefore, harvesting of large trees for timber can have detrimental impacts on the nesting of stingless bees.

In addition to the above groups, other arthropods have also been studied in a habitat degradation context. An examination of the effects of logging and forest conversion to plantation on dung beetle fauna was undertaken in Sabah (Danum Valley Conservation Area) (Davis *et al.* 2001). Dung beetles are critically important components of tropical biodiversity because, being decomposers, they are involved in nutrient recycling. Further, they act as vectors for seed dispersal and, by removing dung, may control the spread of parasites to vertebrates. Although primary forest and its surrounding regenerating forests had similar species-richness of dung beetles, plantations

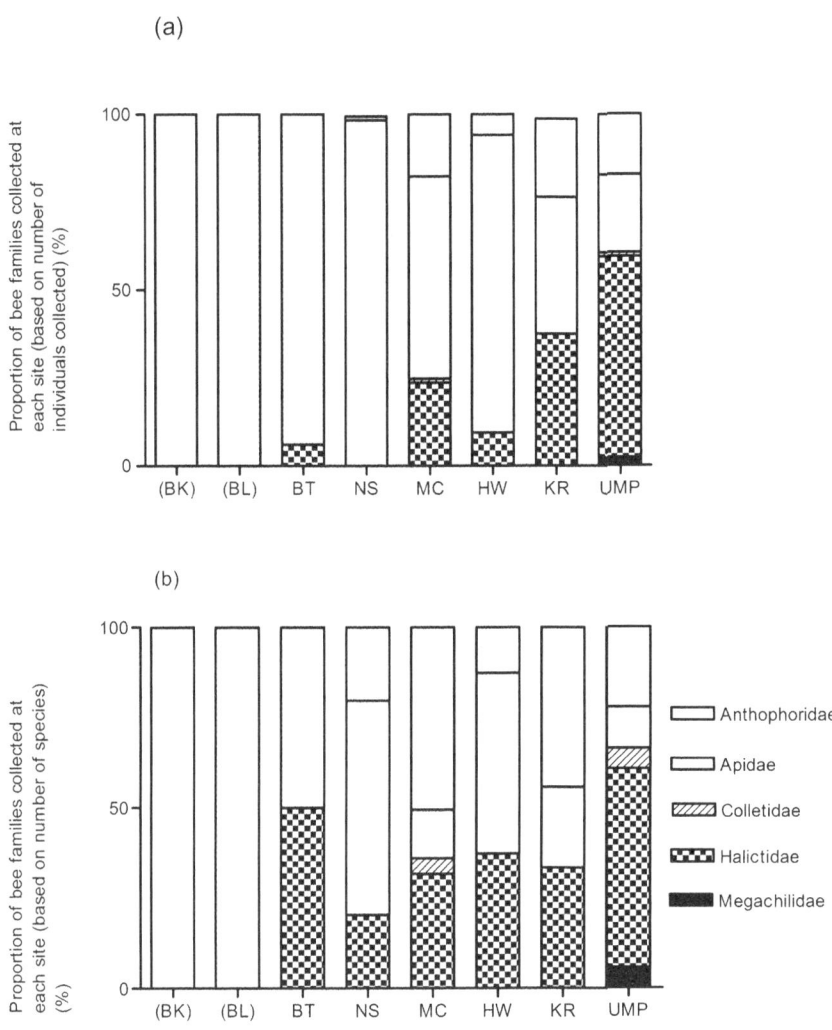

Figure 3.6 (a) Proportion of bee families collected at each site, based on the number of individuals collected. Proportion of bee families collected at each site, based on the number of species collected. BK = Bekok*; BL = Belumut*; BT = Bukit Timah Nature Reserve; NS = Nee Soon; MC = MacRitchie; HW = Holland Woods; KR = Kent Ridge; UMP = UMP oil palm plantations. *Sites in brackets are in Peninsular Malaysia. (Reprinted with permission from Liow *et al.* 2001.)

composed largely of genera such as *Acacia* and *Albizia* had less than half to two-thirds of their species-richness (Davis *et al.* 2001). Dung beetle species associated with the primary forest were most negatively impacted by the conversion of their habitat to plantations.

Leaf litter ants are also considered to be useful indicators of ecosystem disturbance, as they constitute about half of all macroscopic invertebrates in leaf litter of tropical forests. A comparison was made of the leaf litter ant fauna of a contiguous forest with that of two nearby forest fragments (4290 and 146 ha) in Sabah (Danum Valley Conservation Area) (Bruhl *et al.* 2003). Species numbers and diversity in forest fragments reached only 47% of that of the contiguous forest, and species density was also lower. Considering that one of the fragments was over 4000 ha in size, it is possible that considerably larger forests are required for the preservation of natural levels of litter ant diversity.

The effects of anthropogenic disturbance on arboreal ants in Sabah (Mt. Kinabalu National Park) was determined (Floren *et al.* 2001). They compared the arboreal ant fauna of a primary forest with regenerating forests (5–40 years old). Primary forest contained more ant species than the regenerating forests. There was apparently some recovery in ant richness with forest age, but 28 species were restricted exclusively to the primary forest. Further, ant abundances were lowest in the younger regenerating forest (Floren & Linsenmair 2001), and ant communities seemed to be complex and stochastic in the primary forest but deterministic and simple in the disturbed forest. This study shows that arboreal ant faunal richness, as well as community dynamics, is impacted by human disturbance. From the same forests, using all arboreal arthropod fauna, Floren & Linsenmair (2001) found that all species were typically rare in the primary forest and that the relative proportion of ants was lowest in the younger forest (21%). This decline coincided with an increase in Lepidoptera larvae (1–9%). The high predation pressure by ants on other insects may be one of the factors responsible for this pattern.

Reptiles and amphibians

Over the past few decades, in many locations worldwide, amphibian populations have undergone remarkable declines, with some species becoming extinct and many more facing the prospect of extirpation (Beebee 1992; Blaustein & Kiesecker 2002; Collins & Storfer 2003). Factors such as deforestation, draining of wetlands and pollution are often attributed to these population declines. However, more recently, amphibian population declines have been reported from protected and apparently pristine areas such as national parks and nature reserves (Houlahan *et al.* 2000). It is hypothesised that these declines may be mediated by variables such as increased ultraviolet radiation, acid rain and disease. In light of this dangerous predicament, it is particularly disturbing that so few scientific data are available on the amphibian populations of Southeast Asia.

Two undisturbed lowland rain forests of Borneo were sampled over time (Voris & Inger 1995). Three streams of a forest in Sarawak (Malaysia) were sampled in 1962, 1970 and 1984, while two streams in a forest in Sabah were sampled in 1986, 1989 and 1990. Voris & Inger (1995) did not observe any systematic decline in amphibians at these sites over up to 22 years, although individual species showed increases, decreases or steady numbers over the study periods. Lack of detailed knowledge of the biology of the species precluded Voris & Inger (1995) from making any clear hypotheses about the cause(s) of population fluctuations in individual species. However, it was clear from this study that there was no overall, systematic decline in amphibian populations at the study sites. Agreeing with this trend, Brook et al. (2003) found that only 7% of the historically recorded 27 amphibian species vanished locally from Singapore after massive deforestation since 1819. Similarly, only 5% of the historically recorded 123 species of reptiles have been extirpated from Singapore. These results suggest that amphibians and reptiles may be less vulnerable to habitat loss because they require relatively less habitat than larger vertebrates to persist. However, one should be cautious in making such a conclusion, as many extirpations might have gone unrecorded due to the cryptic nature of these organisms. The lack of redundancy in nature reserves in Singapore make existing amphibians and reptiles particularly vulnerable (see Chapter 6).

In addition to the decline and loss of local populations and possible species extinctions, one of the most disturbing findings on reptiles has been that two regional crocodile species (*Crocodylus siamensis* and *C. porosus*) may no longer support viable populations throughout most of their range (Platt & Tri 2000; Thorbjarnarson et al. 2000). The fact that these crocodiles were lost from protected and pristine habitats indicates that in addition to habitat destruction, these species, like amphibians, may be affected by factors such as ultraviolet radiation, chemical pollutants, climate change, emerging infectious diseases and direct conflict with humans (including over-harvesting).

One of the consequences of reduction in habitat is that individual population sizes are reduced for the impacted species. Small populations can result in the loss of genetic diversity and an increase in the risk of chance demographic hazards. A loss of genetic diversity can compromise reproductive success through inbreeding depression (expression of deleterious homozygous genes) and loss of genetic diversity, further reducing their ability to adapt to environmental change (Frankham *et al.* 2002). Therefore, there is a need to identify genetically viable populations of endangered species and make plans to preserve those populations. The genetic diversity of the Komodo dragon (*Varanus komodoensis*), the world's largest lizard was determined (Ciofi *et al.* 1999). This unique species is restricted to only five islands in Indonesia, and is considered to be an endangered species (IUCN 2003). The major threat to this species is loss of habitat. Ciofi *et al.* (1999) found that the population on the island of Komodo had high genetic diversity, and recommended that as a consequence, particularly strident efforts should be made to preserve this population. They also noted that isolated populations on other islands (e.g. Gili Motang) had low genetic diversity, therefore these populations may be particularly vulnerable to further habitat loss and stochastic threats (e.g. environmental catastrophes such as fires or severe storms).

Birds

Birds are arguably the best studied group in the context of forest degradation globally, and Southeast Asia is no exception. Unlike arthropods, most bird studies are not from one location and include both regional and community level studies. However, there is a paucity of species-specific bird studies.

Deforestation data and the species-area equation have been used to predict the number of threatened endemic bird species in Southeast Asia (Brooks *et al.* 1997). Subsequently, in a more detailed analysis, the correlation was determined between deforestation and the existence of threatened endemic species in the lowland and montane forests of Southeast Asia (Brooks *et al.* 1999). Broadly, more threatened bird species were found in heavily deforested areas. However, there were some regional and habitat related differences. The number of threatened species in montane areas was underestimated by the species-area equation, possibly because their restricted ranges made them disproportionately more vulnerable to habitat loss (Brooks *et al.* 1999). Conversely, lowland avifauna of areas such as the Lesser Sundas and Java were less threatened than predicted by the same equations. Because of centuries of habitat loss, these areas may have either lost their more sensitive endemics before any scientific surveys were conducted, or they may simply contain more inherently tolerant species due to their more dynamic environments. Nevertheless, the analyses of Brooks *et al.* (1999) demonstrate that deforested areas generally harbour more threatened species, either because of habitat loss and degradation, or due to associated causes (e.g. hunting). A recent review concluded that of the 274 resident bird species of the Sundaic region (excluding Palawan), 30%

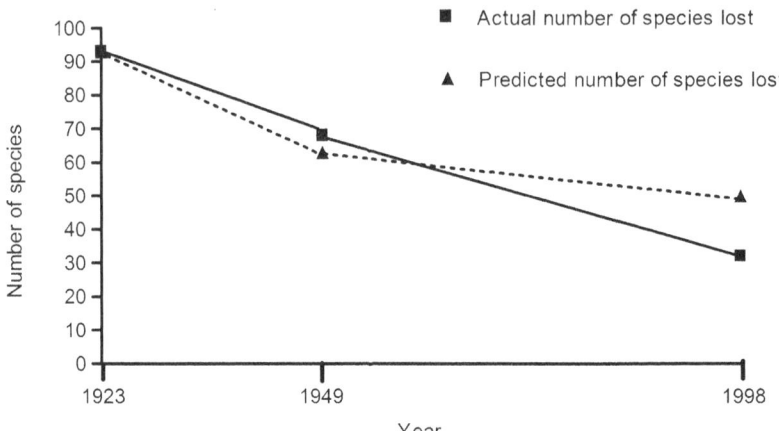

Figure 3.7 Predicted and observed loss of forest birds in Singapore since 1923. Predicted numbers were calculated based on a theoretical species-area equation and assuming that forest clearing started in 1819 in Singapore. (Reprinted with permission from Castelletta *et al.* 2000.)

were affected negatively by fragmentation, and 10% by logging (Lambert & Collar 2002). This finding, along with studies by Brooks *et al.* (1997; 1999), provide a dire indication of the likelihood of extirpation or endangerment of a large proportion of Southeast Asia's forest birds in light of heavy deforestation.

Many empirical studies underpin the grim predictions provided by the above studies. For instance, Singapore has lost 65 of 203 bird species since 1819, with 61 (67%) of the 91 original forest species being extirpated (Castelletta *et al.* 2000; Brook *et al.* 2003). Extinctions have been particularly rapid immediately subsequent to habitat loss, followed by a gradual attenuation in the extinction rate over time (Fig. 3.7).

With regard to the remaining habitat of Singapore, larger patches contain more extant bird species, with only 20 species now restricted to secondary and primary forests (Fig. 3.8; Castelletta *et al.* 2005).

A comparative study was undertaken of various characteristics (e.g. sex ratio, rarity and ectoparasite prevalence) of birds in two small forest fragments in Singapore (483 and 937 ha; Nee Soon and MacRitchie, respectively) with those in two large (2230 and 4196 ha; Matang Wildlife Sanctuary and Gunung Gading National Park) continuous forests of Sarawak (Malaysia) (Sodhi 2002). On average, more bird species and individuals were caught in the mist nets in continuous forests than in fragments. For all the species combined, bird recaptures were at least 18% higher in one of the continuous forests than other sites. The higher recaptures may indicate

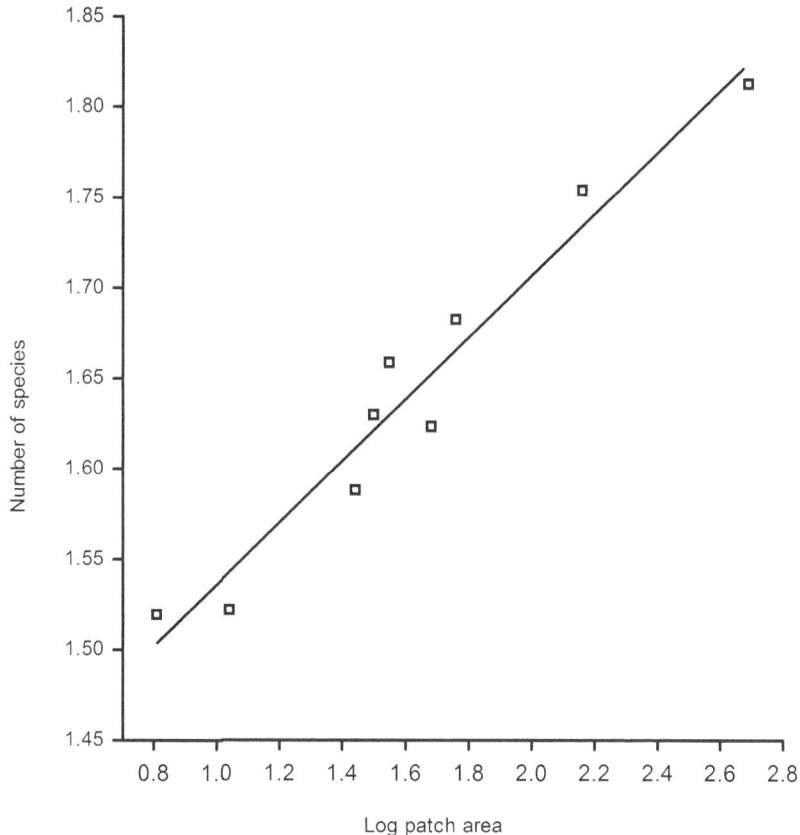

Figure 3.8 Species-area relationship of residents in the Singaporean patches
of old and young secondary forests (M. Castelletta & N. S. Sodhi,
unpublished data, 2000).

lower mortality or greater site fidelity. This site also had at least 9% more rare
species than other sites. Other variables, such as proportion of adults, sex
ratio, and ectoparasite prevalence and intensity, did not differ among the
sites. This is not altogether surprising, because Singapore has lost a large
proportion of its forest birds and may now contain mostly resilient species.
Sodhi (2002) concluded that there may be long-term resilience in some forest
bird species, but that large tracts of undisturbed forests may be needed to
adequately protect rare forest birds.

Habitat fragmentation can lead to an elevation in predation in forest
fragments by making bird nests more accessible to 'forest avoiding' generalist predators such as the house crow (*Corvus splendens*) which typify

disturbed areas. In fact, high nest predation may have been one of the drivers of the extirpation of forest birds from Singapore. However, because actual predation events are difficult to observe, artificial nest experiments have been used to compare predation pressure among sites. In Singapore, 80% ($n = 328$ nests) of artificial ground nests were depredated. One primary forest experienced at least 12% less predation than other fragments (Wong et al. 1998). Some 62% of arboreal nests ($n = 110$) were predated (Sodhi et al. 2003). However, there was no difference in predation pressure among the forest types, and other studies have reported similar predation rates in unlogged and logged forests (albeit on artificial nests). For example, in Peninsular Malaysia (Pasoh Forest Reserve), after five days, in unlogged forest interior sites at least one more nest survived than in logged forest interior sites or in forest edge sites about 35 years old (Cooper & Francis 1998).

In addition to Singapore, there have been studies on the effects of habitat loss and disturbance on birds in Java. Only 11 lowland bird species persisted in the Bogor Botanic Gardens (86 ha) after 50 years of isolation from nearby forests (Diamond et al. 1987). This number represents only about 10% of the lowland avifauna of Java. Diamond et al. (1987) speculated that increased human-mediated disturbance (e.g. nest predation by domestic dogs, Canis familiaris) might have been one of the factors in the extinction of ground-dwelling species such as the banded pitta (Pitta guajana) and black-capped babbler (Pellorneum capistratum). The parasitic brush cuckoo (Cacomantis variolosus) also disappeared from this site, possibly because of the decline in the abundance of its likely hosts, the pied fantail (Rhipidura javanica) and hill blue flycatcher (Cyornis banyumas). In contrast, the parasitic plaintive cuckoo (C. merulinus) survived in this forest isolate because of the continued high abundance of its primary host, the ashy tailorbird (Orthotomus ruficeps) (Diamond et al. 1987).

In central Java (Linggoasri), a survey was undertaken of regenerating rain forest and pine plantation sites in an area where all of the lowland forests had been logged three years prior to the study (Sodhi et al. 2005b). Comparing a bird list for the site prior to logging (van Balen 1999) they found that ten lowland bird species were likely to have vanished following logging. Only 33% of Javan lowland bird species occurred in their study area. However, as with Singapore, the persisting lowland bird species did not appear to be heavily impacted physiologically by habitat degradation, showing no significant loss of body condition or reduction in survival and reproduction. In general, only 10% ($n = 105$) of the arboreal artificial nests were predated in this area, suggesting a relatively low predator density (Sodhi et al. 2003). Predation was 20% less in the selectively logged forest than in secondary forest, but the same proportion (3%) of nests were predated in the selectively logged forest and pine plantations. No small mammal species (potential predators of bird nests) were caught in the pine plantations, one of the possible reasons for the low predation pressure there.

A survey of raptors was undertaken from five forest reserves in Java (Thiollay & Meyburg 1988). These reserves ranged in size between 530 and 50 000 ha. Although the frequency of raptor observations increased with area of reserves, six species (including the Indian black eagle, *Ictinaetus malayensis*) were also discovered in degraded areas such as plantations. This finding implies that these six species may be less vulnerable to forest fragmentation and loss because they can utilise non-pristine habitat. However, the endangered Javan hawk eagle (*Spizaetus bartelsi*) and chestnut-bellied hawk eagle (*Hieraaetus kienerii*) had home ranges of 2000 to 3000 ha suggesting smaller reserves may be inadequate for them. Further, the 8–10 pairs of these species in reserves may not be viable (i.e. having sufficient reproductive success to achieve population replacement) over the long term. Thiollay & Meyburg (1988) concluded that reserves as large as 30 000 ha may be needed in Java to adequately protect existing raptors.

Some studies have addressed the question of how birds are affected by logging by comparing bird faunas of logged and unlogged areas. It has been argued that results from such studies should be viewed with caution as topographical and habitat differences can bias the observed patterns (Jones *et al.* 2003). Therefore, it would be better to sample a site both before and after logging (e.g. Johns 1987). However, such study designs can suffer from problems of pseudoreplication (lack of independent data points), and must be analysed carefully as repeated measures to avoid confounding statistical analyses. Increasing the number of replicates in studies comparing bird communities of unlogged and logged areas is desirable, but not always possible due to heavy deforestation restricting the number of suitable sites.

Surveys of birds and mammals in various forest types (undisturbed, 3–5 years after selective logging and unlogged mature forest) were carried out in east Kalimantan (PT International Timber Corporation, Indonesia concession) (Wilson & Johns 1982). They found that similar numbers of hornbill and pheasant species were found in the selectively logged forest and primary forest, but the densities were considerably lower in the former. Groups of these species also decreased in their area of occupancy immediately after logging. This study suggests that logged forests can be re-colonised by birds provided these areas lie close to a population source (e.g. primary forest). It remains unclear, however, whether these re-colonising populations breed at adequate levels to maintain viable populations.

A comparative study was made of the understorey birds of the Pasoh Forest Reserve with a nearby forest that was selectively logged 25 years prior to the study (Wong 1985). A total of 83 species were mist-netted in primary forest compared to 73 in the logged forest. More captures, perhaps indicating high abundance, were made in primary forest than in the logged forest (1422 versus 948). Food resources (e.g. flowers, fruits and arthropods) were also less abundant in logged than in primary forest (Wong 1986). There was no difference in the proportion of rare species in the two forest sites. The logged forest had a lower proportion of breeding individuals and consequently had fewer juveniles compared to the unlogged forest. Wong (1985) concluded that there was some avifaunal recovery underway in the logged forest, but it was far from complete. Wong (1985) also recommended that logged patches should be smaller and located relatively close to primary forests, so that re-colonisation by species displaced during logging is maximised.

The bird fauna was compared between logged and unlogged forests in Sabah (near Ulu Segema Forest Reserve) (Johns 1996). In general, as shown elsewhere, more bird species were found in logged than unlogged primary forest. However, 15 bird species (including the spotted fantail, *Rhipidura perlata*) were less abundant in logged than unlogged areas. Sighting rate of vulnerable families (e.g. Phasianidae) did not differ between those logged forests distant ($\geq$4.5 km) and close ($\leq$1.5 km) to primary forests ($>$500 ha). This may indicate that re-colonisation of logged areas is largely independent of the source area. However, this inference should be viewed with caution, because different species of a family were pooled. Colonisation abilities can be species- rather than family-specific. Earlier results from the same site (Lambert 1992) found that 29 species (including the chestnut-naped fork-tail, *Enicurus ruficapillus*) were confined to the primary forest. Trogons, woodpeckers, wren-babblers and flycatchers appeared to be particularly negatively impacted by selective logging, being absent or at lower abundance in logged than primary forest.

Earlier studies from Peninsular Malaysia (Tekam Forest Reserve), reported that the richness and diversity of avian families such as Alcedinidae, Trogonidae, Timaliidae, Muscicapidae and Dicaeidae were low in 0–6 year old logged sites compared to unlogged primary forest (Johns 1986, 1989). However, 12 years after logging, the recovery was largely complete, with 181 species recorded in the logged areas compared to 193 in the primary forest. Some 17 of 22 species (including the great hornbill, *Buceros bicornis*), identified previously as intolerant to logging, were not observed in regenerating forests of logged areas. In addition the logged forests still did not reach the species-abundance patterns characteristic of the primary forest (i.e. more low abundance species).

Another comparative study of the bird fauna of unlogged and logged forest was undertaken on the Indonesian island of Seram (Marsden 1998). Although bird diversity was high in unlogged forest, most species occurred in logged forests. Species with low densities in unlogged forest were not highly vulnerable to logging. This is unexpected as rare species, due to their low population density, tend to be particularly susceptible to disturbance (Goerck 1997). Further, species with small global ranges also did not seem to be particularly impacted by logging. However, several endemic forms (e.g. golden bulbul, *Ixos affinis affinis*) seemed to be vulnerable to logging. Therefore, the likely impact of logging on endemic forms (those in most critical need of preservation given considerations of global species conservation), needs to be particularly considered when allocating areas for logging.

A comparison of the bird community of a disturbed mature rain forest (including human activities such as hunting, collection of fruits and seeds, and small-scale logging) with that in a nearby secondary forest (30–40 years old) and a regenerating clearing (5 years old) was carried out in Thailand (Khao Nur Chuchi) (Round & Brockelman 1998). This site is the last known stronghold of the highly endangered Gurney's pitta (*Pitta gurneyi*). The number of bird species in the mature forest, secondary forest and clearing were 110, 109 and 67, respectively. While the secondary forest supported the populations of the endangered Gurney's pitta, specialist frugivores (e.g. green broadbill, *Calyptomena viridis*), woodpeckers (e.g. banded woodpecker, *Picus miniaceus*) and some babblers (e.g. scaly-crowned babbler, *Malacopteron cinereum*) were in low abundance. Round & Brockelman (1998) recommended that secondary forests found along the boundaries of parks and sanctuaries should be actively preserved for the added protection of endangered birds, as they provide both an ecological buffer around the most important primary forests, and reasonable (if suboptimal) habitat in themselves.

Research in the Sungei Tekam Forestry Concession (Peninsular Malaysia), determined the effects of logging on the spectacular hornbills (Johns 1987). Hornbills are good indicator species as their food resources (fruits) are patchily distributed and it is estimated that 500 000 ha of unfragmented forest is required to support a long-term minimum viable population of 10 000 individuals. Further, hornbills require tall trees (mean height $= 38\,\mathrm{m}$) to nest (Marsden & Jones 1997), and these may be available only in primary or late secondary forests. Earlier in a three day survey in Kalimantan (Sapaku River, Rengang Road and Kutai Reserve), researchers found that the helmeted hornbill (*Rhinoplax vigil*) and rhinoceros hornbill (*Buceros rhinoceros*) were more abundant in a primary than in a logged forest (Wilson & Wilson 1975). Johns (1987) showed that although primary

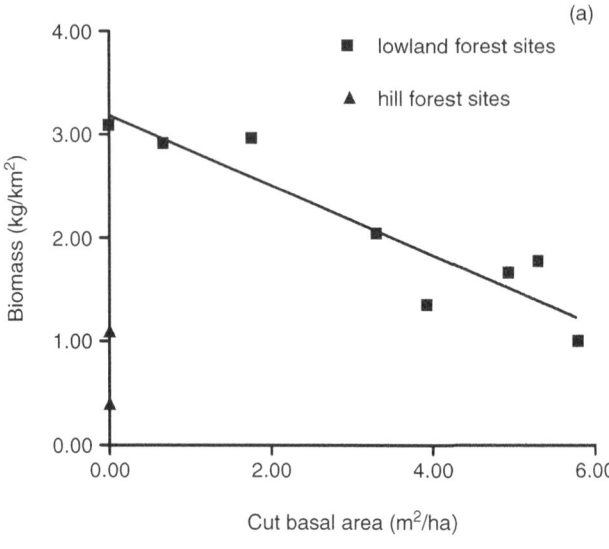

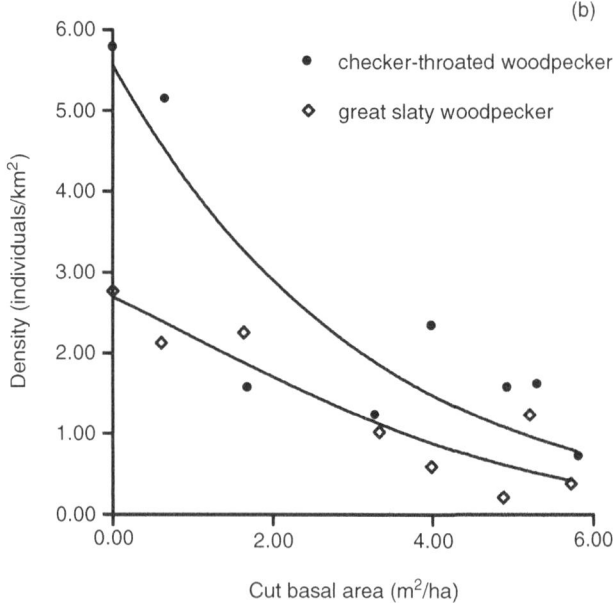

Figure 3.9 (a) Relationship between cut basal area and woodpecker biomass in eight lowland and two hill forest sites in west Kalimantan. Regression for lowland sites: $y = 3.174 - 0.339x$, $r = -0.94$, $R^2 = 87.7\%$, $p = 0.001$. (b) Relationships between cut basal area and densities of checker-throated woodpecker and great slaty woodpecker in lowland forests in west Kalimantan.

forests had higher species-richness and abundance, and logged forests contained 56% fewer foraging trees, most species were nevertheless still present in logged areas. Management actions such as not removing foraging tree species (e.g. *Xylopia* spp.) and trees with cavities suitable for nesting hornbills would certainly enhance the habitat value of selectively logged areas for hornbills. In addition, efforts should be made to ensure that hornbills are not hunted in logged areas, considering that logged forests become more accessible to poachers due to the creation of logging trails and roads. It should be realised that for maximum conservation benefits (i.e. high numbers of breeding hornbills), in addition to other management actions, forests should be allowed sufficient time to regenerate, which for the hornbill habitat could take 80 years or more (Johns 1988).

As in the case of hornbills, woodpeckers are considered particularly sensitive to forest disturbance, because they nest in cavities located in large trees and forage on bark-boring insects of both living and dead large trees. A comparative study of the woodpecker fauna was undertaken in primary forest of the Pasoh Forest Reserve and surrounding secondary forest (40 years old) (Styring & Ickes 2001). A total of 12 species were found in the primary forest and 11 in the secondary forest, with nine woodpecker species common to both forest types. Three of the woodpecker species were more common in the primary forest, whilst one was more common in the secondary forest. This result suggests that whilst overall similar species-richness was comparable in both the primary and regenerating forests, some species flourish only in the primary forest, a trend which is likely to be due to a lack of suitable foraging and nesting trees (dead and dying trees) in the regenerating areas (Styring & Ickes 2003; Styring & Hussin 2004).

The impact of logging of lowland (<70 m elevation) forests on the woodpeckers was studied in west Kalimantan (Lammertink 2004). Woodpecker biomass declined with the quantity of timber (cut basal area) removed from the lowland sites (Fig. 3.9a). Two species, the checker-throated woodpecker (*Picus mentalis*) and great slaty woodpecker (*Mulleripicus pulverulentus*) showed very similar trends, with significant declines in density that were tightly correlated with an increase in cut basal area (Fig. 3.9b). Unlogged hill forests (120–400 m elevation) did not

Figure 3.9 (*cont.*)
Regression for checker-throated woodpecker: $y = 5.743 - 2.036vx$, $r = -0.90$, $R^2 = 80.8\%$, $p = 0.002$. Regression for great slaty woodpecker: $y = 3.008 - 1.043vx$, $r = -0.90$, $R^2 = 80.6\%$, $p = 0.003$. See Lammertink (2004) for details. (Reprinted with permission from Lammertink 2004.)

seem to be optimal habitat for the lowland woodpeckers, as these areas contained even lower woodpecker densities than the heavily logged lowland sites. In addition, the great slaty woodpecker (a species affected by logging), was entirely absent from these higher elevation. Clearly, conservation of reasonable areas of unlogged lowland forests is needed to safeguard the existing lowland woodpecker community (Lammertink 2004).

What proportion of native forest bird richness exists in plantations? Research to compare the bird fauna of primary forests with that of plantations (e.g. rubber, *Hevea brasiliensis*) was carried out in southwestern Sumatra (Indonesia) (Thiollay 1995). Species-richness and diversity was 12–62% less in plantations than in primary forests. Further, community similarity was low (0.43–0.55) between primary forests and plantations. Less than half of the forest species in Sumatra were recorded in any of the three plantations surveyed. Thiollay (1995) suggested that vegetation structure and volume may not be conducive for forest birds in plantations, because of low plant species diversity and lack of suitable food sources. Clearly (as one might intuitively expect) plantations cannot replace primary forests in terms of supporting bird diversity.

However, somewhat contrary research recorded 60% of primary forest species in *Albizia* plantations in Sabah (Sabah Softwoods) (Mitra & Sheldon 1993). Even more surprisingly, the frequency of bird observations (indicative of abundance) was almost twice as high in seven year old plantations than in a nearby primary forest. It was speculated that fast growth of trees in these plantations, and heavy load of associated insect pests, may have resulted in a good habitat for many bird species. However, many birds in these plantations were apparently transient and merely visiting from the nearby primary forest. Plantations were depauperate in woodpeckers, flycatchers and large canopy frugivores. Perhaps the availability of nesting trees and some food types (e.g. fruits) may be limiting these more specialist bird guilds from plantations. This would suggest that the conservation value of plantations can be most effectively enhanced by retaining nearby primary forests. However, one has to be cautious that bird detectability may be higher in plantations than in more densely vegetated rain forests.

In addition to logging, human settlements are altering forested landscapes. These countryside (mixed rural) landscapes contain villages, home gardens and swathes of agricultural areas, often embedded within the forests. Researchers determined the persistence of primary lowland birds in selectively logged forests (30 years old) and countryside in Peninsular Malaysia (Peh *et al.* 2005). They found that although the regenerating forests contained about three-quarters of the primary forest species, the countryside supported only one-quarter (Table 3.1).

Table 3.1 *Bird species-richness (S), absolute numbers of birds observed (N), Shannon's diversity indices (H'), evenness indices (J') and measures of dominance (D) at each study site*

Sites	S	N	H'	J'	D
Bekok					
SBP	157	846	4.35	1.98	0.02
SBS (overall species)	145	823	4.30	1.99	0.02
SBS (primary forest species)	127	763	4.16	1.98	0.03
SBC (overall species)	85	903	3.63	1.88	0.04
SBC (primary forest species)	51	249	3.27	1.92	0.06
Belumut					
GBP	148	779	4.39	2.02	0.02
GBS (overall species)	140	737	4.22	1.97	0.03
GBS (primary forest species)	123	654	4.14	1.98	0.03
GBC (overall species)	74	1139	3.29	1.76	0.06
GBC (primary forest species)	42	227	2.91	1.79	0.10

For Bekok, SBP = Primary forest; SBS = Secondary forest; SBC = Agricultural land. For Belumut, GBP = Primary forest; GBS = Secondary forest; GBC = Agricultural land (with permission from Peh *et al.* 2005).

Further, the avifauna of the disturbed areas was dominated numerically by a few common species. Although some forest birds can carry out breeding activities in logged forests (Fig. 3.10), continuous primary forests appear to be critical for sustaining a high diversity of extant forest bird species, especially in providing a sufficient variety of food sources.

Similarly, the value of countryside locales for birds was studied in central Sulawesi (Lore Lindu National Park) (Sodhi *et al.* 2005c). The impact of ongoing deforestation on the highly endemic birds of Sulawesi (which constitute 33% of the avifaunal diversity) is poorly known. Mixed-rural habitat (e.g. villages, roads, gardens and scattered forest remnants) and secondary forest (40 years old) contained 82% and 72%, respectively, of 34 primary forest species. However, 15 of these supported the highest abundances in primary forests. Similar results have been reported by another study from the same area (Waltert *et al.* 2004). Simulation modeling by Sodhi *et al.* (2005c) suggested that forest species that need forest cover require at least 20% continuous tree cover to have a high likelihood of persisting in mixed-rural areas, with a very strong threshold decline in habitat suitability when tree cover falls below 15% (Fig. 3.11).

The effects of fragmentation on Southeast Asian forest bird species-richness has received surprisingly scant attention in the literature. Van

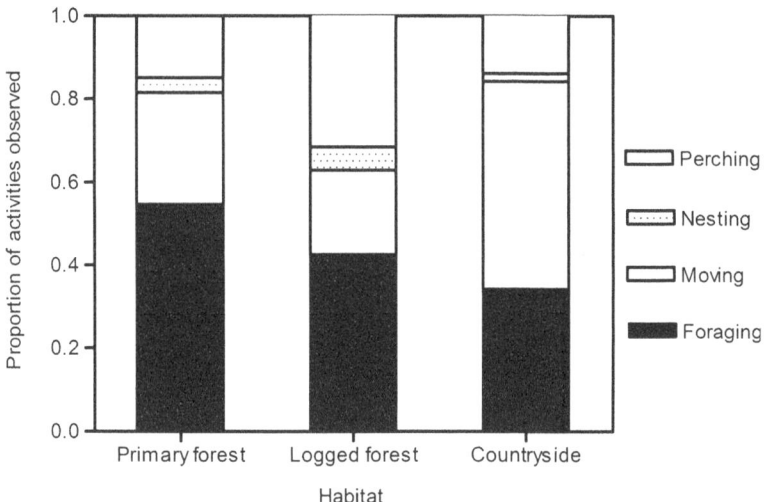

Figure 3.10 Proportion of bird individuals observed engaging in different behavioural activities (Reprinted from Peh *et al.* 2005). With permission from Elsevier.

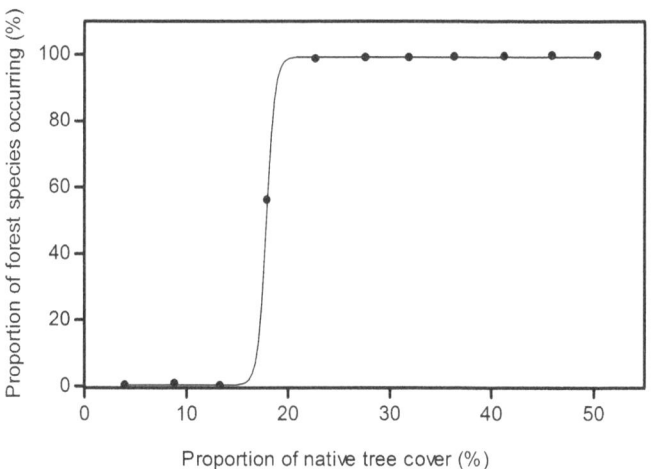

Figure 3.11 The proportion of forest species occurring in simulated samples with an increasing proportion of native tree cover (Reprinted from Sodhi *et al.* 2005). With permission from Elsevier.

Balen (1999) sampled birds of 19 lowland forests (6–50 000 ha) on the island of Java. He found that 30 forest species (i.e. those found in forest edge and forest interior habitats) occurred only in forests >1000 ha in area (Table 3.2), and 13 of these were restricted to forests >10 000 ha.

Table 3.2 *Javan lowland forest birds most at risk from fragmentation*

Patch size	Species		No. of forest patches
Large (>20 000 ha)	Buff-rumped woodpecker	*Meiglyptes tristis*	0
	Scaly-breasted bulbul	*Pycnonotus squamatus*	0
	Yellow-eared spiderhunter	*Arachnothera chrysogenys*	0
	Greater flameback	*Chrysocolaptes lucidus*	1
	Long-billed spiderhunter	*Arachnothera robusta*	1
	Violet cuckoo	*Chrysococcyx xanthorhynchus*	1
	Blue-banded kingfisher	*Alcedo euryzona*	1
	Thick-billed flowerpecker	*Dicaeum agile*	1
Medium (10 000–20 000 ha)	Large green pigeon	*Treron capellei*	1
	Orange-backed woodpecker	*Reinwardtipicus validus*	1
	Asian paradise-flycatcher	*Terpsiphone paradisi*	2
	Maroon-breasted philentoma	*Philentoma velatum*	2
	Crimson-breasted flowerpecker	*Prionochilus percussus*	2
Small (<10 000 ha)	Crimson-winged woodpecker	*Picus puniceus*	3
	White-bellied woodpecker	*Dryocopus javensis*	4
	Blue whistling thrush	*Myophonus caeruleus*	3
	Banded woodpecker	*Picus miniaceus*	3
	Crimson sunbird	*Aethopyga siparaja*	3
	Fulvous-chested jungle flycatcher	*Rhinomyias olivacea*	3
	Yellow-vented flowerpecker	*Dicaeum chrysorrheum*	3
	Silver-rumped swift	*Rhaphidura leucopygialis*	4
	Rufous woodpecker	*Celeus brachyurus*	2
	Checker-throated woodpecker	*Picus mentalis*	3
	Dark-throated oriole	*Oriolus xanthonotus*	4
	Green imperial pigeon	*Ducula aenea*	2
	Asian fairy bluebird	*Irena puella*	4
	Cream-vented bulbul	*Pycnonotus simplex*	5(? +)
	Great slaty woodpecker	*Mulleripicus pulverulentus*	5
	Malaysian cuckooshrike	*Coracina javensis*	3
	Lesser cuckooshrike	*Coracina fimbriata*	5

Modified from van Balen (1999).

This result suggests that many forest species in Java may require large patches of intact habitat to survive. Additionally, many of these species seem to occur in very few locations, and may thus be particularly vulnerable to environmental stochasticity and disturbance.

The negative impacts of fragmentation have also been documented for birds and mammals from two wildlife sanctuaries in Thailand (Pattanavibool & Dearden 2002). Both wildlife sanctuaries (Om Koi and Mae Tuen) contained a mixture of montane and evergreen forests. Patches in Om Koi were larger (>400 ha) and connected whereas those in Mae Tuen were smaller (<100 ha) and isolated. Overall, Om Koi contained 119 bird species compared to 89 in Mae Tuen. In Mae Tuen, almost half of the species were found at low abundances (<1 individual/visit), but this predicament was true for only one-third of the species at Om Koi. Large frugivores such as the brown hornbill (*Ptilolaemus tickelli*) and great hornbill (*Buceros bicornis*) were only found in Om Koi. This study illustrates the general rule of thumb in conservation ecology that protection of larger, connected patches is better for maintaining bird species-richness. The same conclusion was reached in Singapore after sampling birds from 17 patches of varying sizes (7–935 ha) (Castelletta *et al.* 2005).

Very little is known about the conservation value of fragments over the long-term. The faunal change over 100 years (1898–1998) in a 4 ha patch of rain forest in Singapore (Singapore Botanic Gardens) has been studied (Sodhi *et al.* 2005a). Over this period, many forest species (e.g. green broadbill) were lost, and replaced with introduced species such as the house crow. By 1998, 20% of individuals observed were introduced species, with more native species expected to be extirpated from the site in the future through competition and predation. This study shows that small fragments decline significantly in their value for forest birds over time, and are vulnerable to competition from invasive species.

There has been a concern that studies reporting the impacts of tropical deforestation and degradation on bird species need to be better designed (Sodhi *et al.* 2004a; see Chapter 6). We admit that rapid deforestation in many Southeast Asian countries has limited the number of studies with sufficiently robust sample sizes and many replicates. However, the effects of habitat alterations on individual bird species are poorly documented and should receive considerably more attention in future.

Mammals

Unlike the case for birds, the majority of studies reporting the effects of habitat degradation on mammals in Southeast Asia have been based on single, charismatic species. As a consequence, we are better able to understand the effects of habitat degradation on individual species in particular circumstances, but are hard pressed to draw community-level generalities. Below we summarise the results from various studies.

The orangutans (*Pongo pygmaeus* and *P. abelii*) are arguably one of the taxa most often used to attract conservation dollars, and are in a sense emblematic flagship species for the conservation movement's goals. It is perhaps indicative of the biodiversity crisis then that this genus, now threatened, was once widespread in Asia but is now restricted to the large islands of Borneo and Sumatra, each island harbouring a different species. Orangutans are threatened due to habitat loss and degradation, hunting for bush meat, and capture of young animals for the pet trade (Robertson & van Schaik 2001). Orangutans are typically found in lowland forests and freshwater and peat swamp forests. Studies show that if given a choice, this species avoids logged areas (Rao & van Schaik 1997). Surveys of orangutans (*P. pymaeus*) were carried out in different types of peat swamp forests in central Kalimantan (Sungai Sebangau Catchment) (Morrogh-Bernard *et al.* 2003). They showed that the densities of sleeping platforms (nests) was highest (636 nests/km^2) in tall interior peat swamp forest in central Kalimantan. Further, recently disturbed areas (subject to timber extraction within the previous two years) had nest densities of orangutans four times lower than those in forests that were disturbed more than two years prior to the study. Similarly, unlogged peat swamp forest in west Kalimantan (Gunung Palung National Park) had 21% higher nesting density of orangutans than logged areas (Felton *et al.* 2003). The density of orangutans in logged forest in Sumatra (Gunung Leuser National Park) was 40% less than that in unlogged forest (Rao & van Schaik 1997). These results suggest strongly that selective logging does impact negatively on these important species.

Due primarily to habitat loss, orangutan populations declined by 45% over a six year period in the Leuser Ecosystem of Sumatra (van Schaik *et al.* 2001; Fig. 3.12). Orangutan groups may have local traditions, primarily in tool use, which are maintained through social transmission. Local extinctions, hunting pressure and habitat loss may all be responsible for the loss of these local traditions in orangutans (van Schaik 2002). Habitat disturbance seems also to affect other behavioural traits of orangutans. In some logged areas, orangutans' dietary preferences shifted from frugivory to folivory, and they travelled more and rested less than in unlogged areas (Rao & van Schaik 1997). This was partly because fruit abundance was 40% lower in the logged than in unlogged areas. Other studies also demonstrate a positive correlation between food abundance and the density of orangutans (Buij *et al.* 2002). In fact, if fruits are readily available in logged areas, orangutan density in these areas can be comparable to that in primary forests (Knop *et al.* 2004).

How large a reserve area do orangutans need for survival? Considering their high site fidelity and large home range size (850 ha for adult females and 2500 ha for adult males), Singleton & van Schaik (2001) found that large

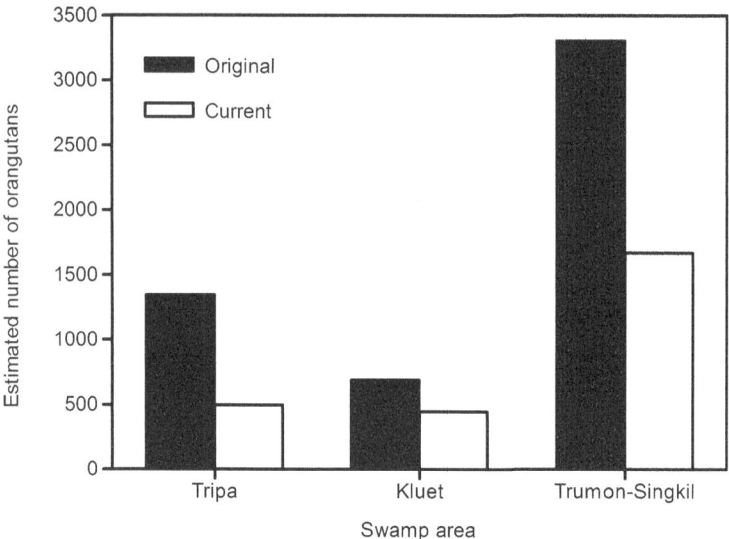

Figure 3.12 Changes in the estimated number of orangutans inhabiting the three swamps of Aeh Selatan, 1993–8. (Data from van Schaik *et al.* 2001.)

tracts in the order of $500\,\mathrm{km}^2$ are needed to support long-term viable orangutan populations. Buij *et al.* (2002) recommended that large tracts of primary forests encompassing varying altitudes may be necessary to preserve the remnant orangutan populations. The above studies have shown that habitat degradation can significantly affect orangutan populations and behaviour, and that they have a large area requirement.

Civets play an important ecological role in the forest as they disperse seeds and keep rodent populations in check. The ranging behaviour of the Malay civet (*Viverra tangalunga*) was compared in a logged and unlogged forest in Sabah (Ulu Segama Forest Reserve) (Colón 2002). It was found that the density of civets was 57% ($2.17/\mathrm{km}^2$ versus $0.93/\mathrm{km}^2$) higher in the unlogged than in logged forests. Similar results were also found from an earlier study from the same area (Heydon & Bulloh 1996). They found that the abundance of eight species of civets was higher in unlogged than in logged forest. Overall, there were 31.5 civets/km^2 and 6.4 civets/km^2 in unlogged and logged forests, respectively. Neither the range sizes nor the activity levels of the civets differed between the forests for malay civets (Colón 2002). These results led Colón (2002) to conclude that the behaviour of this species is not seriously impacted by selective logging, and recommended that unlogged forests be preserved close to logged areas so that there is the possibility that civets from the former can colonise the latter.

However, from the same area in Sabah, different results were obtained for two mouse-deer species (*Tragulus javanicus klossi* and *T. napu borneanus*) (Heydon & Bulloh 1997). The densities were measured of both these frugivorous ungulates in selectively logged (2–12 years old) and primary forest. Both species were less common in the logged areas. Densities of both species correlated positively with food resources. Heydon & Bulloh (1997) concluded that dietary inflexibility (preference for easily and faster digestible fruits) in these two mouse-deer species may preclude them from exploiting logged areas.

Gibbons (*Hylobates* spp.) are important and wide ranging seed dispersers in Southeast Asian forests. In Kalimantan (Barito Ulu Research Area), gibbons disperse about 80% of the seeds from the fruits that they consume (McConkey 2000). The estimated effective dispersal rate was 13 seedlings/ha/group/year. In Kalimantan (Tahura Bukit Soeharto), the impact of fragmentation on the Bornean gibbon (*H. muelleri*) was studied, surveying seven isolated patches (16–25 ha) (Oka *et al.* 2000). These forests were probably fragmented during the 1960s. Oka *et al.* (2000) noted that roads (4–6 m wide) may limit the dispersal of gibbons among the patches, and could eventually result in detrimental effects associated with long-term isolation, including inbreeding and loss of genetic diversity.

The proboscis monkey (*Nasalis larvatus*) is endemic to the island of Borneo. This species, like orangutans, is threatened by habitat loss, hunting and exploitation for the illegal pet trade. This species was eliminated from the Pulau Kaget Nature Reserve (Kalimantan) because illegal agricultural expansion pushed this species to the fringes of the reserves where individuals began to starve (Meijaard & Nijman 2000). This led to the remaining animals being translocated to a nearby unprotected area and to a zoo. Of 145 translocated animals, 34% died due to poor post-translocation management. This study shows that management of a species should be carefully planned if such disastrous results are to be avoided. Like the proboscis monkey, the Sulawesi crested black macaque (*Macaca nigra*) also is threatened by habitat loss and hunting. As the name indicates, this species is endemic to the Sulawesi region. Population surveys of Sulawesi crested black macaques were conducted in two undisturbed and one disturbed forest on the islands of Sulawesi and Bacan (Rosenbaum *et al.* 1998). The highest density of this species was found in the primary forest (170 individuals/km^2), where higher quality food resources (e.g. ripe fruits) may exist compared to logged forests, resulting in a higher carrying capacity for this species in the undisturbed habitat.

Some native Southeast Asian mammal species can actually benefit from habitat loss and the associated demise of predators. Exceptionally high densities of native wild pigs (*Sus scrofa*) (27–47 pigs/km^2) were found in the Pasoh Forest Reserve (Peninsular Malaysia) (Ickes *et al.* 2001). This was attributed to the local extinction of natural predators (e.g. tigers, *Panthera tigris*) and the year-round food supply of African oil palm fruits (*Elaeis guineensis*). It is likely that similar trends could arise for some other groups of animals, such as birds.

In addition to single-species studies, the community approach has been applied to document the effects of habitat degradation on mammals. Brooks *et al.* (1999) used deforestation data and the standard species-area equation ($z = 0.25$) to predict the number of threatened endemic mammal species in Southeast Asia. There were more threatened mammals in the montane regions of Southeast Asia than predicted by this technique. For example, there were four times more threatened mammals in montane areas of the Philippines than predicted. Factors such as hunting pressure compound the effects of deforestation on threatened Southeast Asian mammals (Brooks *et al.* 1999).

Wilson & Johns (1982) compared mammals of different forest types (e.g. logged, plantations versus primary forest) in Kalimantan (PT International Timber Corporation Concession). They found that out of eight mammal species, only one (Malayan sun bear, *Helarctos malayanus*) was restricted to the primary forest. Densities of all mammal species were generally higher in the primary than in other forests, again indicating the importance of this forest type. Good management practices such as maintenance of large trees with cavities, protection of figs (*Ficus* spp.) and elimination of poaching has also been shown to facilitate the persistence of sun bears in logged areas (Wong *et al.* 2004).

Johns & Johns (1995) determined the recovery of primates after 18 years of selective logging (50% tree loss) in Peninsular Malaysia (Sungai Tekam Forestry Concession). Logging resulted in high mortality of juveniles; there was recovery in juvenile numbers in the white-handed gibbon (*Hylobates lar*) but not in banded leaf monkey (*Presbytis femoralis*) nor the dusky leaf monkey (*Trachypithecus obscurus*) (Fig. 3.13a).

These results suggest that for leaf monkeys, fewer juveniles would reach maturity, and thus group size will be lower, in logged forests. Supporting this hypothesis, the mean troop size for leaf monkeys was substantially lower in logged forest than in their unlogged counterparts, although this was not the case for gibbons (Fig. 3.13b). It is unclear what impact this reduction in group size might have on the biology of leaf monkeys in logged forests. Another concern for these primates, as mentioned earlier, is that due to increased access enjoyed by poachers, illegal hunting can become rampant in logged forests (Johns & Johns 1995).

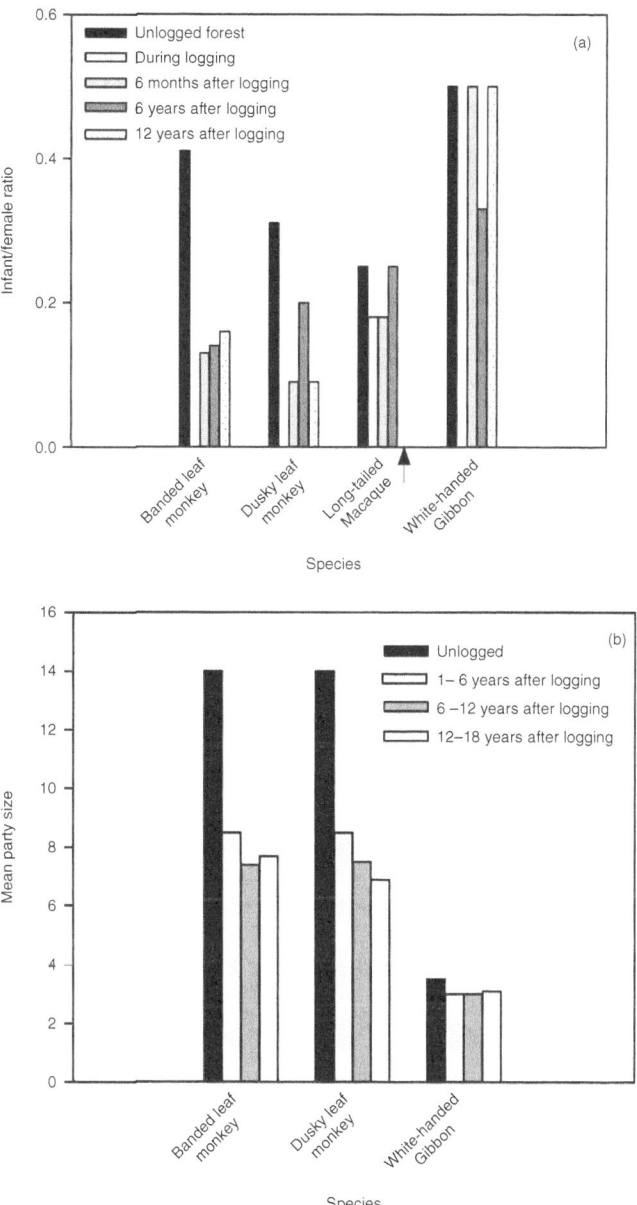

Figure 3.13 (a) Estimated infant/female ratios among primates at Tekam Forest Reserve. (b) Mean party (group) sizes for surveyed primates in Tekam Forest Reserve. (Data from Johns & Johns 1995.) (In (a) there was no data for long-tailed macaques, 12 years after logging.)

A comparative study was made of the mammal species-richness of seven protected areas (70–304 ha) and their adjoining logged forests in Peninsular Malaysia (states of Pahang and Selangor) (Laidlaw 2000). The size of the natural forest area was the most important variable affecting mammal richness, with a sharp fall observed from 164 to 70 ha of natural forest. Laidlaw concluded that the preservation of undisturbed forest adjacent to disturbed areas can effectively enhance mammal richness. However, large continuous forest may be needed for the survival of the largest carnivores and herbivores, such as the Asiatic elephant (*Elephas maximus*). Further, even within very large areas of continuous forest (e.g. the 27 469 km^2 Taman Negara National Park, Peninsular Malaysia, and contiguous forests in southern Thailand) some of the larger mammals, including the tiger (*Panthera tigris*) may avoid areas with high human traffic (Kawanishi & Sunquist 2004). Similar results have been found by Pattanavibool & Dearden (2002) who compared the mammal richness of two contiguous wildlife sanctuaries in Thailand. With larger fragment sizes and greater connectivity (see above), Om Koi had ten more mammal species than Mae Tuen, and only Om Koi retained large mammals like the Asiatic elephant. This result suggests that fragmentation impacts negatively upon mammal richness in these study areas.

Small mammals have important roles in forest ecosystems such as pollinators, seed dispersal agents and food for carnivores. In Peninsular Malaysia (Bangi Forest Reserve and Krau Wildlife Reserve), more species of small mammals were found in a primary than a regenerating forest selectively logged 18 years prior to the study (17 versus 11) (Akbar & Ariffin 1997). The density of small mammals was also higher in primary than in regenerating forest (27.8 versus 23.6 individuals/ha). Similar results have been reported by other studies from Malaysia (Harrison 1969; Zubaid & Rizal 1995; Yasuda *et al.* 2003).

A meta-analysis of the impact of deforestation on the Southeast Asian biota

An increasingly common means of summarising and combining the results of multiple studies in ecology is through the use of statistical 'meta-analysis'. Meta-analysis, in the broadest sense, is a scientific review in which the emphasis is placed on quantitative synthesis of data, using a range of available techniques of measurement and analysis to evaluate effect sizes for a phenomenon of interest across different studies (Mann 1990). It can be particularly useful for identifying the presence and strength of results from a group of studies, even when there are apparently contradictory observations with the group (Fernandez-Duque & Valeggia 1994). Below we describe a meta-analysis of the literature on the impact of deforestation and habitat degradation on the biota of Southeast Asia, taken from the results of the studies described in the previous sections of this chapter (B. W. Brook & N. S. Sodhi, unpublished data, 2004).

We collated from a total of 84 individual publications from Southeast Asia in which ecological attributes had been recorded in pristine and nearby deforested/disturbed sites, with a total of 720 pairwise comparisons reported across five broad taxonomic groups, plants, arthropods, reptiles, birds and mammals, six broad measures of 'ecological health', Richness, Abundance, Structure, Diversity, Demography and Others and a range of different impact types (clear-cut or selective logging, fire, conversion for agriculture, etc.). Overall almost 70% of pairwise comparisons supported this expectation, with mammals being the most sensitive group (82%) and invertebrates the least (62%), and measures of species-richness (76%) and abundance (71%) being the most informative measures of impact.

Extinction-proneness related to life history and ecological traits

Some species are more susceptible to habitat disturbance than others due to intrinsic characteristics such as large size, naturally restricted distributions and behavioural habitat specialisation. For instance, restricted range butterfly species seem particularly vulnerable to changes in their habitat (Hill *et al.* 1995; Hamer *et al.* 1997; Hill *et al.* 2001; Ghazoul 2002; Hamer *et al.* 2003). In Vietnam (Tam Dao Mountains), endemic butterflies were confined to mature montane forest (Spitzer *et al.* 1993), and within these forests, the understorey supported more endemic butterfly species than the canopy. This result illustrates the high conservation value of mature montane forests for butterflies, and the surprisingly crucial role played by the understorey vegetation in supporting native butterfly biodiversity.

Specialisation may also cause a species to be particularly sensitive to disturbance. In Singapore, for example, those butterfly species which avoided non-forested habitats (e.g. urban parks) were more likely to be forest and larval host plant specialists than those commonly found in urban environments (Koh & Sodhi 2004). Larval host plant specificity was also found to be the single most important correlate of extinction risk for butterflies in Singapore (Koh *et al.* 2004a). This finding is consistent with the theory that specialised species with narrow ecological niches are less adaptable to rapidly changing environmental conditions, such as those associated with habitat loss and degradation (McKinney 1997; Purvis *et al.* 2000). Previous studies from temperate regions have reported similar trends, with larval host plant specificity being an important factor in the survival and distribution of butterflies from these areas (Thomas 1991; Thomas & Morris 1994; Shahabuddin et al. 2000; Thomas et al. 2001; Shapiro 2002).

Forest dependent taxa, as expected, have been shown to be acutely vulnerable to extinction following deforestation and forest fragmentation. Using various taxonomic groups (vascular plants, phasmids, butterflies, decapods, freshwater fish, amphibians, reptiles, birds and mammals), Brook et al. (2003) showed that in Singapore, forest taxa (those residing predominantly in primary or secondary rain forest or mangrove forest interiors) lost 33% species compared to 7% species losses for species that prefer or tolerate open or forest-edge habitats. A probable explanation is that forest-dependent species are more likely to decline with the loss of forest cover (due to reductions in breeding and feeding sites, increased predation, elevated soil erosion and nutrient loss, dispersal limitation, enhanced edge effects, etc.) and thus suffer higher extinctions than non-forest-dependent species that are better able to persist in disturbed land-scapes (e.g. urban areas).

Forest birds provide the best quantitative data on the selectivity of extinctions. In Singapore, substantially more extinctions occurred among forest birds (61 of 91 species) than non-forest-dependent birds (13 of 127 species); and among insectivores (76%) compared to other feeding guilds (e.g. carnivore, frugivore or granivore) (Castelletta et al. 2000). Larger birds were more extinction-prone than smaller birds (Castelletta et al. 2000; Brook et al. 2003). Similar results have been found from Java, where large birds, insectivores and frugivores were marked by high extinction-proneness (Sodhi et al. 2005b). Large animals tend to be more extinction-prone, perhaps due to allometric scaling laws (Peters 1983), they have naturally lower population sizes, lower reproductive rates, larger area requirements and/or higher food intake than small animals (Terborgh 1974; Leck 1979; Pimm & Kitching 1988). Specialist frugivores may be vulnerable in degraded or reduced forests because such areas no longer provide a suffi-cient volume of year-round fruit (Leck 1979), and insectivores may be more extinction-prone because of impoverishment of insect fauna, coupled with their inherently poor dispersal abilities (Sekercioglu et al. 2002).

In Sumatra, the bird species not observed to be utilising agroforests (e.g. rubber tree plantations) were predominantly forest-dependent species, and included large insectivores and frugivores (Thiollay 1995), underscoring their susceptibility to the large-scale conversion of forests for agriculture that is now occurring throughout much of Sumatra. In Peninsular Malaysia, of the 159 extant primary forest bird species, Peh et al. (2005)

found only 28–32% existed in disturbed areas (e.g. plantations and country-side). The types of microhabitats on which these Malaysian species depended also seemed to influence their vulnerability to disturbance. For instance, the ground-dwelling species tended not to occur in disturbed areas, presumably because most of the ground-dwelling species nest at or close to the ground. Higher predation of artificial eggs in disturbed compared with undisturbed forests has been observed (Wong *et al.* 1998; Sodhi *et al.* 2003), indicating that ground-dwelling primary forest birds may suffer higher nest predation in disturbed areas.

High investment in sexual traits may render highly sexually dimorphic species less adaptable to changing environment (McLain *et al.* 1999). However, sexual dimorphism was not one of the predictors of the forest bird abundance on the islands of Sumba and Buru (Indonesia) (Jones *et al.* 2001). The role of sexual dimorphism in extinction-proneness remains equi-vocal (Sodhi *et al.* 2004a).

The above studies suggest that the susceptibility of tropical species to habitat disturbance may be related to their ecological traits (Koh *et al.* 2004a), as shown for a range of taxa in other environmental settings (Purvis *et al.* 2000). Identifying such ecological correlates of extinction-proneness could be invaluable for defining conservation priorities, espe-cially in Southeast Asia, where traditional methods of threat assessment, such as the International Union for the Conservation of Nature and Natural Resources (IUCN) Red List criteria (Randrianasolo *et al.* 2002) are either lacking, or do not provide adequate resolution of relative endangerment due to data deficiencies (Koh *et al.* 2004a).

Loss of ecological interactions

It would be simplistic to assume that the death of one species would have no wider ecological ramifications (Koh *et al.* 2004c). The tropical rain forests of Southeast Asia are characterised by highly diverse and complex ecological communities, in which many species are inextricably dependent upon one another. However, conservation biologists have tended to focus on the study of the independent declines or extinctions of individual species, whilst largely ignoring the possible cascading effects of species co-extinctions (e.g. hosts and their parasites) (Stork & Lyal 1993; Koh *et al.* 2004c). A recent study showed that the decline and loss of butterfly species in Singapore was positively associated with the decline and loss of their specific larval host plants (Koh *et al.* 2004b) (Fig. 3.14).

Similarly the parasitic brush cuckoo probably disappeared from the Bogor Botanic Garden in Java, Indonesia, following the decline of its

Figure 3.14 Locally extinct butterfly, *Parantica aspasia* and its locally extinct hostplant, *Tylophora* sp.

hosts (Diamond *et al.* 1987). It is likely that similar co-extinctions between other interdependent taxa have occurred but gone unrecorded in Southeast Asia. Further studies are urgently needed to investigate the phenomenon of species co-extinctions, as it could have important ecological and conservation implications, such as improved understanding of species distribution patterns and the estimation of extinction rates of interdependent taxa (Koh *et al.* 2004b).

Extinction or decline of a species can also disrupt ecological processes, the end result of which may also be cascading co-extinctions. Frugivory, a key interaction linking plant reproduction and dispersal with animal nutrition, is placed in jeopardy by habitat degradation. Research was undertaken to determine whether frugivorous bird assemblages were impacted adversely by forest disturbance (Zakaria & Nordin 1998). They made their observations of species visiting figs (*Ficus* spp.) in primary dipterocarp forest and recently logged forest (2 years prior to the study) in Sabah (Danum Valley Conservation Area). Species composition was similar between the forests, with only one species unique to each site – the jambu fruit dove (*Ptilinopus jambu*) was only found in the primary forest, whilst the olive-winged bulbul (*Pycnonotus plumosus*) only in the logged forest. However, all obligate frugivores (e.g. fairy bluebird, *Irena puela*) had a lower visitation rate to fig trees in logged rather than in primary forest.

Zakaria & Nordin (1998) suggested that there may be less food available for frugivores in logged forest, and that altered microclimatic conditions (e.g. greater light penetration) may make logged forests less conducive sites for foraging frugivores. This may, in turn, result in lower seed dispersal rates in logged areas, further reducing fruit abundance.

There has been concern raised that the declining availability of fruits in disturbed Southeast Asian forests may result in the extirpation of under-storey frugivores (e.g. green broadbill), which have relatively small ranges and are generally unable to tolerate extreme microclimates (e.g. high temperature in the logged forests) (Lambert 1991; Zakaria & Nordin 1998). It is both interesting and disturbing to note that the green broadbill has been driven to local extinction in Singapore following the island's massive loss of forest.

On the island of Negros (Philippines), while early-successional tree species were visited by a wide-spectrum of frugivores, mid- and late-successional trees were largely visited by hornbills and fruit pigeons (Hamann & Curio 1999). Thus mid- and late-successional species were more specialised in their use of seed dispersers. This pattern places them in a highly vulnerable position, as the avian frugivores upon which they rely are endangered, primarily because of high hunting pressure. Clearly, should these avian frugivores be extinguished, mid- and late-successional trees will lose their dispersers and cascading co-extinctions could be the unfortunate end result.

Figs, a keystone resource in Southeast Asian rain forests (Lambert 1991), rely on tiny (1–2 mm) species-specific wasps for pollination. Some fig wasps may have limited dispersal ability. Harrison *et al.* (2003) suggested that forest fragmentation can have a negative influence on these wasps and the figs that they pollinate. Koh *et al.* (2004c) showed that at least 3% (about 8) fig wasp species will be liable to go extinct with the extinction of their host figs.

Mammals can also play a critical role in the regulation of plant populations and communities. For example, the wild pig exhibits several behaviours, such as soil rooting, nest building and seed predation, which can influence the structuring of understorey vegetation. Soil rooting can kill seedlings and facilitate the spread of invasives (Lacki & Lancia 1983; Aplet *et al.* 1991), but conversely, can also promote plant growth (Lacki & Lancia 1983). Wild pigs were superabundant (27–47 pigs/km^2) in Pasoh Forest Reserve (Ickes & Thomas 2003). Here, pigs reduced plant recruitment by three-fold but seemed to increase plant growth by 52% for trees between 1 and 7 m tall (Ickes *et al.* 2001). Therefore, this species clearly influences plant dynamics in the understorey, and its demise (or superabundance) can strongly impact upon forest composition and structure.

Deforestation has ramifications beyond those immediate to the forests. Almost all flowering plants in tropical rain forests are pollinated by animals

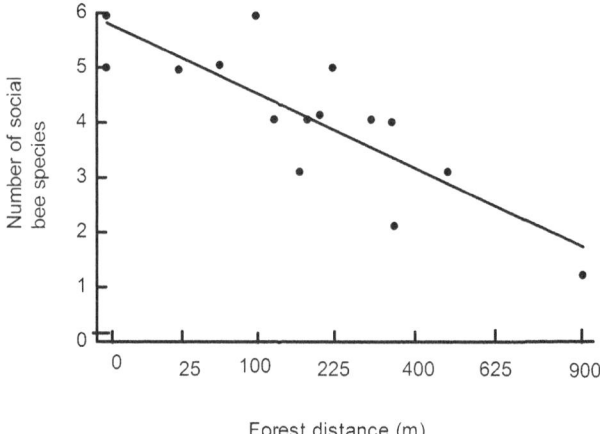

Forest distance (m)

Figure 3.15 Relationship between the number of social bee species and the forest distance: $y = 5.87 - 0.14x^2$, $F = 27.63$, $R^2 = 0.68$, $n = 15$, $p < 0.001$. (Reprinted with permission from Klein *et al.* 2003a.)

(Bawa 1990). It is also estimated that one-third of the human diet in tropical countries is derived from insect-pollinated plants (Crane & Walker 1983). Therefore, a decline of forest-dwelling pollinators may impede plant reproduction not only in the forests, but also in the neighbouring agricultural areas. Lowland coffee (*Coffea canephora*) is an important tropical cash crop, and it depends on bees for cross-pollination. In central Sulawesi (Lore Lindu National Park) the number of social-species bees visiting coffee fields decreased with the increasing distance from the forest (Klein *et al.* 2003a) (Fig. 3.15). Species-rich bee communities increased the pollination success of coffee plants (Klein *et al.* 2003b). Such findings illustrate the imperative of preserving native forests within close proximity to agroforestry systems to facilitate the travel by pollinating insects which depend on the forests for nesting and year-round sources of pollen and nectar.

Loss of specialised behaviours

Clearly, habitat loss and degradation affects both individual species and inter-specific ecological interactions. It can also influence behaviour.

Many species of tropical birds feed in mixed species flocks. Despite their vulnerability to forest disturbances in the Neotropics (Sodhi *et al.* 2004a) the consequences of forest perturbation on mixed species flocks is poorly understood in the threatened Southeast Asian rain forests. Nevertheless, researchers were able to examine the effects of local-scale habitat disturbance on mixed species flocks along an escalating gradient of anthropogenic modification (i.e. forest interior, forest edge and urban) in a submontane tropical rain forest in Peninsular Malaysia (Fraser's Hill) (Lee *et al.* 2005). Mixed species flocks in the forest interior and forest edge habitats had significantly higher numbers of species (9.30 ± 2.30 and 8.35 ± 2.31, respectively) than those observed in the urban habitat (5.07 ± 1.65). Flock participation was influenced by environmental characteristics (e.g. canopy cover), and flocking species sensitive to habitat disturbance were likely to be from families Corvidae, Nectariniidae and Sylviidae, which had restricted altitudinal ranges and were exclusively dependent on primary forest and understorey microhabitat. This study shows that sub-montane mixed species bird flocks are impacted by urbanisation. Undoubtedly, many such specialised behaviours are negatively impacted by deforestation, and more scientific attention is needed in this direction.

Summary

1. The biodiversity of Southeast Asia remains poorly studied relative to other tropical and subtropical areas. More research is needed to evaluate precisely the effect and extent of ongoing massive habitat alterations on Southeast Asian biodiversity. However, the results available to date document the same negative relationships between deforestation and habitat degradations and biodiversity health that has been observed in other tropical and temperate regions.

2. Among biotic groups, vascular plants are, surprisingly, among the least studied in Southeast Asia. More research effort in this critical group is particularly needed.

3. There is a strong and consistent effect of deforestation and habitat degradation on a range of measures of ecological health, including abundance, diversity and species-richness, and consistent impacts across different taxonomic groups.

4. The high conservation value of large primary forests has been repeatedly demonstrated. Studies also show that preserving primary forests in degraded landscapes, coupled with appropriate management practices (e.g. retaining cavity forming trees), can elevate their biodiversity value.

5. Data show that restricted range species, those with specialised behaviours, and forest-dependent species, are particularly impacted by habitat destruction, though this area requires further investigation across a more diverse range of taxa.

6. Extirpation of some species may also influence the survival of closely linked species. For example, the loss of frugivores may have repercussions on seed dispersal, and ultimately forest regeneration. More data to understand (and ultimately prevent) the loss of species of sensitive biotic interactions such as pollination and frugivory, are needed.

Chapter 4

Beyond deforestation: additional threats to Southeast Asian biodiversity

In this chapter, we review the effects of various threats to Southeast Asian biodiversity that are not directly caused by habitat loss and fragmentation. These processes include wildfires, overexploitation of species from hunting and the pet trade, invasive species, climate change and novel zoonotic diseases. Clearly of course, many of these factors are not mutually exclusive to habitat loss, and may be indirectly caused or enhanced by habitat degradation. For example, the creation of trails during logging operations can facilitate hunting by increasing access to the forest interior.

Forest fires

The periodic climatic warming phenomenon known as El Niño can induce severe and extended dry conditions in Southeast Asia and other areas. The combination of El Niño-mediated drought conditions and poor land use practices can facilitate the penetration of forest fires deep into tropical forests that were previously thought to be fire resistant (Uhl 1998; Fig. 4.1). Transmigration of humans into forests and an increase in accessibility due to creation of roads can further exacerbate the spread of forest fires (Stolle *et al.* 2003). Forest fires have always been present in the Southeast Asian landscape, but factors such as rapid human population growth, change in land-use and apparently increasingly severe bouts of El Niño are now working in combination to increase the probability and intensity of catastrophic fires (Kinnaird & O'Brien 1998; Taylor *et al.* 1999).

During 1997–8, Southeast Asia experienced a very dramatic and widespread episode of forest fires, when up to 5 million ha of Indonesian rain forest (in Sumatra and Kalimantan) was burnt (Schweithelm 1998). These fires, unprecedented in scale in modern times, seem to have been caused primarily by poor logging practices, and small- and large-scale land clearing for agriculture and tree plantations. The resulting smoke and ash from these fires blanketed much of Indonesia, Malaysia, Singapore and northern

Figure 4.1 Fire cleared land in the Philippines.

Australia. This smoke not only jeopardised the health of approximately 20 million inhabitants but also disrupted the economies of these nations because of a decline in tourist numbers (Talbott & Brown 1998). It is estimated that the effect of these fires in the worst hit areas was equivalent to each inhabitant smoking fours packs of cigarettes each day (Talbott & Brown 1998). Thus understandably, due to cardiovascular and respiratory complications arising as a result of elevated air pollution levels, these fires seem to have increased the mortality of the older (65–74 years old) inhabitants of Kuala Lumpur (Sastry 2002). In Indonesia alone, 32 000 people suffered respiratory ailments most likely triggered by these fires (Kinnaird et al. 1998). Additionally, the total economic loss to the region from this environmental disaster was estimated to be US$4.4 billion (Kinnaird & O'Brien 1998).

In addition to the effects on the region's human population, these fires impacted directly upon the flora and fauna. It is estimated that 4.6% of the canopy trees died and between 70%–100% of seedlings and 25%–70% of saplings were wiped out in Sumatra (Barisan Selatan National Park) due to the 1997–8 fires. The fires facilitated the influx of exotic plant species such as *Chromolaena odorata* into the area (Kinnaird & O'Brien 1998). However, the full extent of damage caused by exotic plant species on the native flora could not be documented fully. The density of the helmeted hornbill (*Rhinoplax vigil*) declined

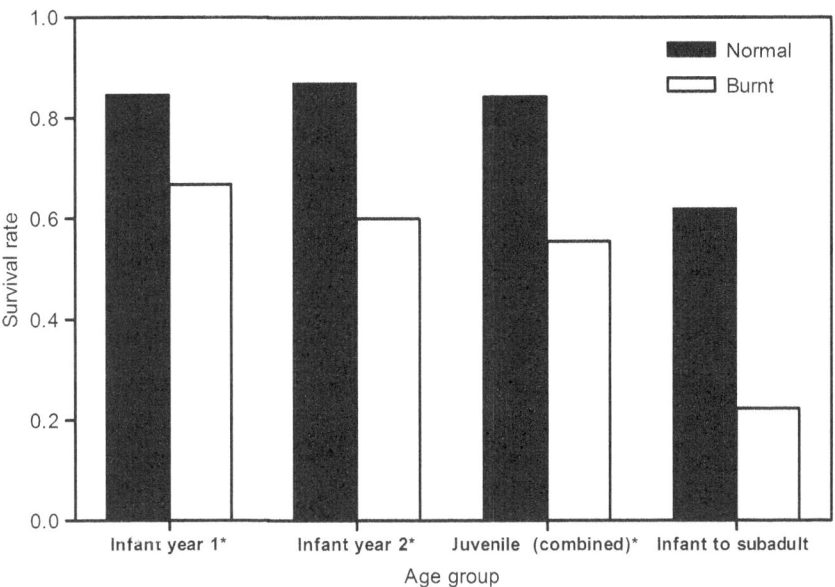

Figure 4.2 Survival rates of siamangs in normal and burnt forests, * indicates statistical significance ($p < 0.05$). (Data from O'Brien *et al.* 2003.)

by 50% after the fires (Kinnaird & O'Brien 1998), though two other hornbill species, the bushy-crested hornbill (*Anorrhinus galerius*) and wreathed hornbill (*Aceros undulates*), seemed largely unaffected, indicating their likely resilience. The loss of flowering and fruiting trees seemed to cause the precipitous decline of many frugivorous species (e.g. barbets, *Megalaima* spp.) in the burnt areas. Siamangs (the largest of the gibbons, *Holobates syndactylus*) disappeared initially from the burnt areas, and groups of banded leaf monkey (*Presbytis melalophus*) were found to be 20% less abundant in the burnt areas than a nearby unburnt forest. Similarly, the horse-tailed squirrel (*Sundasciurus hippurus*), a primary forest specialist, also disappeared from burnt areas (Kinnaird & O'Brien 1998). Factors such as the availability of resources (e.g. fruits and cavity nesting trees) influenced this species' persistence, and subsequent re-colonisation of burnt areas.

From the same area of Sumatra as above, a more detailed study was conducted to determine the longer-term effects of fires on the siamang (O'Brien *et al.* 2003). Siamang groups were significantly smaller in burnt than unburnt areas (3.2 versus 4.0), and the chances of infant siamangs surviving to adulthood was lower in the burnt areas (Fig. 4.2). One of the reasons for this observation could be that the siamang's main food, strangler figs (*Ficus* spp.), were reduced by 48% in burnt compared to unburnt

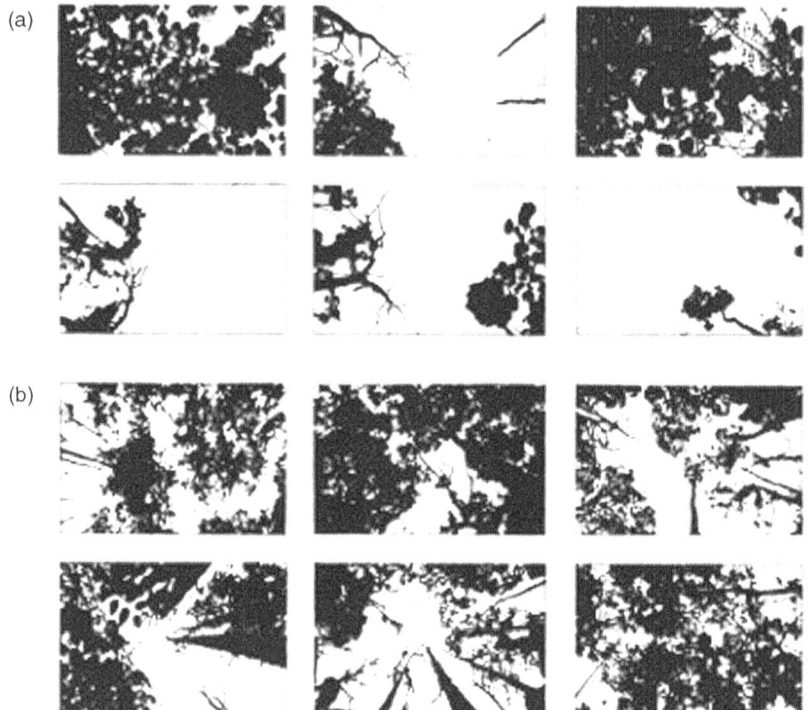

Figure 4.3 Vertical photographs showing the effects of logging on forest canopies. (a) Logged burnt. (b) Unlogged burnt. (Reprinted with permission From Woods 1989. Copyright 1989 Association for Tropical Biology and Conservation.)

areas. O'Brien *et al.* (2003) concluded that the fires have impacted negatively upon this species. Further, this study implies that the decline in the number of seed dispersers such as the siamang may retard vegetation recovery in burnt areas.

There are other studies on the consequences of fires on the native biotas of Southeast Asia. In Sabah (Sandston Hill Dipterocarp Forest, Sepilok Forest Reserve, Labuk Road Forest Reserve and Ulu Tiulon area), tree mortality after drought and fire ranged from 38%–94% and 19%–71% in logged and unlogged (primary) forests, respectively (Woods 1989) and logged forests had more severe canopy loss compared to burnt unlogged forests (Fig. 4.3). However, both drought and fire caused similar mortality (80%) in both types of forests.

Thus, high tree mortality in logged areas following fire may diminish their prospects of vegetation recovery, as the growth of grasses (which, in turn, provide the understorey fuel for future fires) will proceed unabated. On the other hand, lower fire-related tree mortality in unlogged areas elevates their prospects for vegetation recovery (i.e. pristine forest is more resilient), although species composition may be lower in the post fire forest (Woods 1989).

Additional support for the above idea has been provided by comparing the tree species (dbh ≥ 10 cm) composition of primary forests and burnt forests in Kalimantan (between Balikpapan, Mt. Beratus and Samarinda) (Slik *et al.* 2002). There were 22 more species of trees recorded in the primary forest than in a forest one year after fire. Even 15 years after fire, tree diversity (Fisher's α) was at least 30 points lower in burnt areas than in the primary forest. Pioneer genera such as *Macaranga* dominated the canopy of regenerating forest 15 years after burning. Despite these observations, Slik *et al.* (2002) concluded that burnt forests still show signs of a gradual recovery in tree diversity, and therefore retain some conservation value, especially if only lightly burnt. However, repeated burning may thwart this conservation potential, because further reductions to an already depleted seed bank can reduce seriously the sprouting potential of plants in areas that are burnt repeatedly (van Nieuwstadt *et al.* 2001).

In Thailand, forest fires, the majority of which are caused by humans, occur frequently. The effects of forest fires on seed and seedling dynamics in seasonal forest were studied in western Thailand (Mae Klong Watershed Research Station) (Marod *et al.* 2002). Many species such as *Pterocarpus* spp. showed higher germination rates and the emergence of larger seedlings in years with fire than years without fire. There were also cases of high levels of resprouting following fire. Marod *et al.* (2002) hypothesised that in a seasonal forest, many plant species may bet hedge against fire by producing seed and seedling banks, and by favouring life history strategies that permit rapid recovery from damage (e.g. resprouting). However, this may not be the case for tropical rain forest plants that are likely to have evolved to cope with only very sporadic fire episodes over time. On the other hand, the effects of recent forest fires in seasonal forests may be underestimated, because they occur so frequently (i.e. at least once a year) and are deliberately started by people (Corlett 2005). Frequent fires in these drier forests are likely to lead to a reduction in the complexity of the vegetation and suppressed species diversity (Corlett 2005).

As reported above, fires can impact the native fauna (Kinnaird & O'Brien 1998; O'Brien *et al.* 2003). Independent of the Sumatran work, the impact of forest fire on the endemic red-knobbed hornbill (*Aceros cassidix*) in Sulawesi (Tangkoko Nature Reserve) was studied (Cahill & Walker 2000). Success of nests that directly experienced fire was significantly lower (62%)

than in previous non-fire years (around 80%). Normal recruitment for this species has been estimated to be 0.32 fledglings/female/year (Kinnaird & O'Brien 1998), but during the fire year, this value almost halved to 0.17 fledglings/female/year. These results suggest that fire can exert negative effects on the reproductive success of this species that has already been impacted by habitat loss, which is also a major problem in this part of Sulawesi.

Invertebrates are also impacted by fires. El Niño-induced forest fires dramatically altered the butterfly community in Kalimantan (Balikpapan–Samarinda region). Two years after fire, the community was dominated by large-winged generalist species and did not contain any of the endemic species present prior to fire (Cleary & Genner 2004). This study shows that endemic butterfly fauna may be affected, at least over the short-term, by forest fires.

In addition to facilitating forest fires, El Niño-induced drought can disrupt vital ecological processes. In Sarawak (Lambir Hills National Park), such an episode reduced the production of inflorescences by dioecious figs (*Ficus* spp.), and may have led to the local extinction of up to eight species of dependent fig wasps (Agaonidae) (Harrison 2000). Figs and fig wasps are a fascinating example of coevolution. Fig wasps raise their offspring in the secure environment of the fig inflorescence, and in return provide the fig plant with a pollination vector. Thus the extirpation of fig wasps can disrupt the pollination of figs and this may have cascading effects on frugivorous vertebrate species depending upon these for food (Harrison 2000; Koh *et al.* 2004c).

Overexploitation of wildlife

Humans have been hunting wildlife for at least 40 000 years in Southeast Asia (Zuraina 1982; Milner-Gulland & Bennett 2003). However, with the decline in forests and ever increasing human densities, the pressure of hunting on wildlife has recently increased immensely. The situation is exacerbated by factors such as the creation of roads, technological advances in weaponry (e.g. better hunting guns) and the dysfunctional nature of institutions meant to enforce quotas and hunting permits (Bulte & Horan 2002; Ling *et al.* 2002; Smith *et al.* 2003). There is a growing realisation that wildlife overhunting in the tropics is not a hyperbole, and that the ensuing 'bush meat' or 'wild meat' crisis (overhunting of wildlife by humans for consumption) is one of the gravest threats to tropical animal biodiversity. The crisis is driven primarily by the following three factors: 1) high market/subsistence demand for wild meat and associated animal products for traditional medicines and ornamentation, 2) poor enforcement of wildlife protection laws, including inadequate patrolling of protected areas, and 3) poor awareness in local communities towards wildlife laws and plight of their wildlife.

The estimates of annual tropical wildlife harvest remain high but imprecise. Overall, wildlife is extracted from tropical forests at approximately six times the sustainable rate (Bennett 2002). For example, an estimated 23 500 tonnes of wildlife meat are consumed annually in Sarawak alone, with an estimated 2.6 million animals shot there every year (Bennett *et al.* 2000; Bennett 2002). In neighbouring Sabah, an incredible 108 million animals are estimated to be shot every year (Bennett *et al.* 2000). Such figures are clearly alarming, and it is perhaps not surprising that the Bennett *et al.* (2000) study from Borneo showed hunting can depress substantially the densities of exploited rain forest populations of mammals and birds (Fig. 4.4).

Slow-breeding large animals such as apes, large carnivores and elephants are particularly vulnerable to hunting (Robinson *et al.* 1999). The potential for population recovery in these animals over short time scales is low (Brook & Bowman 2004). Supporting this hypothesis, there is evidence that within the past 40 years, primarily due to excessive hunting, 12 large vertebrate species have been extirpated from Vietnam (Milner-Gulland & Bennett 2003). Some of these extinctions can have implications for forest recovery: loss of keystone species such as seed dispersers (e.g. hornbills) can be a detriment to forest regeneration (see Chapter 3).

Commercial logging is a key facilitator of rampant overexploitation of forest animals (see Chapter 1). Such operations increase access to remote areas, bring people from other regions, and change local economies and resource consumption patterns. Bush meat forms a component of up to two-thirds of the meals of highland people in Sarawak (Bennett *et al.* 2000). Most of the wildlife hunting in Sarawak occurs along the roads opened up by logging (Robinson *et al.* 1999). Commercial logging companies can, however, help to alleviate wildlife hunting, being often the only formal institutions in remote areas. 'Green labelling' and independent party certification can be used as positive incentives for forest managers to encourage logging companies to support environmentally-friendly practices such as ensuring that their staff do not hunt or bring native wildlife meat into logging camps (Robinson *et al.* 1999).

The overexploitation of wildlife not only jeopardises the animal populations that fall prey to this pressure, but also the survival and cultural heritage of the people who rely on them (Bennett *et al.* 2000). The most vulnerable consumers remain traditional forest peoples. Although the 'bush meat crisis' has galvanised biologists, the solutions may be difficult, but not

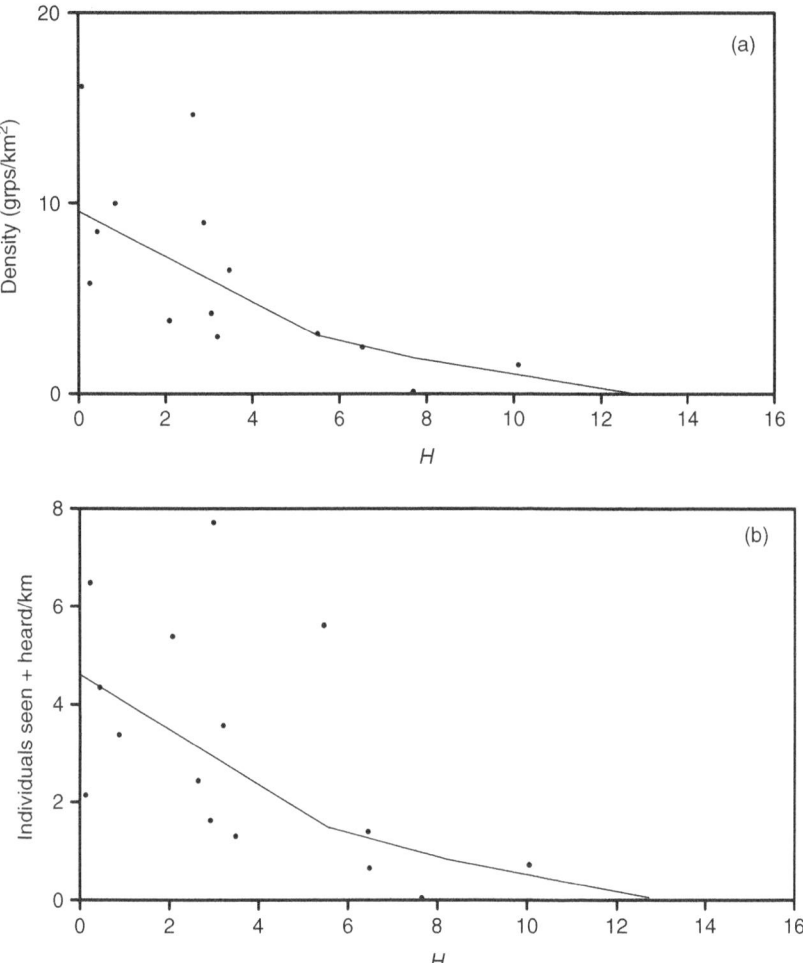

Figure 4.4 Effects of hunting on animal densities. (a) Primate density
and (b) Number of hornbill sightings plus calls per km of survey walked,
versus a hunting pressure index (*H*) in study sites throughout Sabah and
Sarawak (From Bennett *et al.* 2000).

impossible, to implement (Robinson & Bennett 2002). A study from
Peninsular Malaysia (Lima Belas Estate Forest Reserve) where a 76 ha
lowland rain forest in the midst of oil palm plantations was protected
from hunting retained an extremely rich primary forest bird and mammal
fauna (Bennett & Caldecott 1981). However, a large-scale total ban of
wildlife hunting in the tropics is neither possible nor feasible (Rowcliffe

2002), but there have been calls to regulate its utilisation to ensure sustainability (Ling *et al.* 2002). As hunting in the tropical forests is largely illegal and not capitalised or industrialised, implementing regulatory mechanisms will of course, be difficult. Further, most tropical countries lack governmental institutions or political will to manage hunting activities. Therefore imposing hunting quotas, regulation of hunting (e.g. ages and sexes of animals to be hunted) and education of hunters will be hard (Robinson *et al.* 1999; but see below).

There is broad consensus that scientific understanding needs to underpin policies for bush meat hunting (Bennett *et al.* 2000; McGowan & Garson 2002; Rowcliffe 2002). However, studies on the biological as well as social angles of the bush meat crisis have been few (Bennett *et al.* 2000). There is a need to develop models on sustainable harvesting (i.e. ensuring the combination of hunting plus natural mortality does not exceed population recruitment) for the key species, even in situations where knowledge of a given species' ecology and demography is imperfect (e.g. Brook & Whitehead 2004; also see below). In addition, broad scale analyses that pinpoint the local and regional 'hotspots' of the bush meat crisis are needed so that immediate actions can be taken where they are most urgently needed (Fa *et al.* 2002).

Development agencies need to take the bush meat crisis into account so that poverty alleviation programmes are not mutually exclusive from the health and long-term well being of the wildlife populations upon which the indigenous people depend (Davis 2002). It should also be borne in mind that bush meat can sometimes pose a serious health risk to humans if precautions in food preparation and animal handling are inappropriate. As a rather extreme but significant example, bush meat consumption of simians in Africa has been posited as a potential source of HIV (Human Immunodeficiency Virus causing Acquired Immune Deficiency Syndrome [AIDS]) origins, and will be likely to continue to expose humans to yet new viruses and resulting diseases (Wolfe *et al.* 2004).

A multi-pronged approach is what is most needed to alleviate the bush meat crisis, with long-term solutions requiring the tandem consideration of economic, social and biological issues (Bennett *et al.* 2000; Pratt *et al.* 2004). A tangible approach should include efforts to: 1) provide alternative income and protein sources for people who rely on wildlife for their everyday subsistence needs (also see below), 2) curtail or at least mitigate the wildlife trade, 3) better protect the 'protected areas', and 4) educate the hunters and buyers on the risks associated with overexploitation of wildlife resources (Robinson & Bennett 2002; Milner-Gulland & Bennett 2003). Sarawak provides a heartening example where some of these measures are currently being implemented to alleviate bush meat hunting (Bennett *et al.* 2000).

Figure 4.5 (a) A pair of nesting maleos (see http://www.maleo.nl. Rights owned by Alain Compost). (b) Maleo nest excavation. (c) Fish as an alternative food to maleo eggs for local people.

Below we identify some of the major impacts of bush meat hunting on Southeast Asian fauna. Galliformes (including grouse, pheasants, partridges and quails) are particularly vulnerable to hunting because they are predominantly ground-nesting and can occur at high densities in certain locations (McGowan & Garson 2002). About one-quarter of all galliformes are now listed as threatened by IUCN, and 90% have over-hunting as one of the listed threats (McGowan & Garson 2002). To highlight the plight of galliformes, we provide an example of the maleo (*Macrocephalon maleo*).

Endemic to the islands of Sulawesi and Buton (Indonesia), the charismatic maleo faces an imminent risk of extinction. Weighing about 1.6 kg, the maleo is a terrestrial megapode (Fig. 4.5a) that has a unique nesting behaviour. As many as 100 pairs nest communally at traditional sites around geothermally heated soils in forests or in solar-heated sandy beaches or riverbeds. Using their powerful feet, adult birds excavate their individual 60–70 cm deep burrows (Fig. 4.5b).

Each burrow contains one egg, with the depth at which it is laid dependent upon the surrounding soil temperature, which ranges between 30 and 35° C. Eggs are covered with sand and are left to incubate via environmental sources of heat. Adults do not provide any parental care. Young are able to dig their way to the surface and can feed independently almost immediately after emergence. Maleos feed on insects, seeds and legumes on the ground in lowland forests (<1200 m elevation).

Breeding between October and April, a pair of maleos can lay up to 12 eggs in 1 year (Prawiradilaga 1997). At about 250 g each and with a high yolk content, maleo eggs (16% of a female's body mass) are five times heavier than chicken eggs. This concentration of highly nutritious protein has long attracted humans to maleo nesting grounds. Initially, local people used to collect maleo eggs for a traditional ceremony. This situation has now changed, as maleo eggs are considered both delicious and nutritious throughout Indonesia. In some local markets, maleo eggs are sold between Rupiah 2500 and 5000 (<US$0.60). Despite being listed on the CITES

(Convention on International Trade in Endangered Species of Wild Fauna and Flora) Appendix I (a list of species threatened with extinction whose international trade is prohibited) and protected under Indonesian law, uncontrolled egg-collecting by the burgeoning human population of Sulawesi and Buton has led to maleo population declines and abandonment of many former nesting sites. Of 131 known nesting sites, 42 have been abandoned in recent times, with another 38 severely threatened, 12 of unknown status and only 5 (4%) not yet considered threatened (Butchart & Baker 2000). The global population has been currently estimated to be between 4000 and 7000 breeding maleo pairs. However, this population is declining rapidly, with 90% declines in some areas since the 1950s (Argeloo & Dekker 1996).

In addition to heavy human exploitation, maleos are also losing foraging habitats due to extensive deforestation of their native islands (Whitten *et al.* 1987a; see also Chapters 1 and 3). Clearly, the future of this species is in extreme jeopardy. There is a dire need to better understand the ecology (e.g. recruitment and site fidelity) of this unique bird, and the status of this species needs to be evaluated critically throughout its range with data collected over multiple years, in order to identify precisely those areas experiencing population declines. To halt the further population decline of this species, immediate actions are needed. Although over 50% of the known maleo nesting sites are protected (Dekker 1990), such protection is inadequate and poorly enforced. There are numerous maleo captive breed-ing programmes in central Sulawesi (Christy 2002) but their success will be limited without the assurance that the released birds have a legitimate chance of surviving to reproduce in the wild. To achieve tangible conserva-tion outcomes, social aspects need to be integrated fully into conservation plans. For instance, there is an urgent need to develop and promote sustain-able egg-harvesting regimes with the full involvement of the local residents. Local communities should be assisted in exploiting alternative food sources, as has been done for a community in Pakuli (central Sulawesi). Here, locals are encouraged to rely on a fish (*Osteochilus* sp.) so as to reduce their harvest of maleo eggs (see Fig. 4.5c). This community is a custodian of their local maleo nesting grounds. In our minds, similar approaches adopted across Sulawesi will work best to ensure the survival of maleos, and other exploited species more generally. Transition to alternative prey, however, is difficult at least by some local communities (Bennett *et al.* 2000).

Swiftlets in Southeast Asia provide another high profile case of over-exploitation (Fig. 4.6). Three swiftlet species are exploited commercially in Southeast Asia: the edible-nest swiftlet (*Collocalia fuciphaga*), Germain's swiftlet (*C. germani*) and the black-nest swiftlet (*C. maxima*). These birds

Figure 4.6 Edible-nest swiftlet (http://www.orientalbirdimages.org).
(Reprinted with permission of R. Surachai, Bangkok.)

nest inland and in coastal caves. Male swiftlets weave nests from thin
gelatinous strands secreted from their salivary glands. These strands are
formed into a half-cup that bonds the nest to the inside of a cave wall. For
centuries, the Chinese have eaten the nests of swiftlets as a delicacy for soups
or in jelly mixed with spices or sweets. It is believed traditionally that the
nests contain aphrodisiac properties. Chemically, however, swiftlet nests
have been made primarily of proteins (50–60%) (Kong *et al.* 1987). While
there is glycoprotein in them that can promote cell division in the immune
system, it may be lost during the cleaning process prior to eating, meaning
that the nests probably have minimal medicinal properties. Nevertheless,
the demand for swiftlet nests has soared over the past 30 years (Sodhi & Er
2000). Today, a top-quality swiftlet nest can sell for US$2500–4000/kg. The
biggest national supplier of swiftlet nests is Indonesia, with other exporting
countries including Thailand, Vietnam, Singapore and Myanmar. To satisfy
the increasing international demand for this wildlife product, Indonesians
began to farm swiftlets in the early 1990s. Swiftlet farmers buy old houses
with colonies of uniform swiftlets (*C. venikorensis*), and use them to cross
foster the eggs of edible-nest swiftlet. It is claimed that one-third of swiftlets
exported from Indonesia now come from farmed swiftlets, but this claim is

difficult to verify. There is also the danger that inter-species hybridisation in swiftlet farming will have spillover effects to wild populations.

Nest collection is a traditional occupation in many Southeast Asian communities. However, because of its high price, swiftlet nest collection is also a lucrative avenue of income for poachers. Primarily because of over-exploitation, swiftlet populations are now under intense pressure. While concrete data are lacking, studies suggest that many swiftlet populations have been severely depleted, such as those in the Niah Caves of Sarawak (Er *et al.* 1995). As local people in some communities rely upon swiftlet nest harvesting, declining yields have also been a cause of concern from a socio-economic angle (Valli & Summers 1990). There have been some studies which attempt to evaluate quantitatively the question of how nest harvests might reduce reproductive success and subsequent recruitment (Kang *et al.* 1991; Tompkins 1999). Good harvest management can certainly result in stable or increasing populations of swiftlets, as illustrated in a study from south-central Vietnam (Casellini *et al.* 1999). Here, strict, sustainable nest-harvesting by a state-owned company has resulted in an increase in nest production of 3% per annum (Casellini *et al.* 1999). The nests are harvested in two phases. The first phase occurs when 10–15% of nests have eggs, as early harvesting means that most pairs can build a new nest. The second phase occurs after 160 days, when almost all nestlings have fledged. This is an excellent example of sustainable harvesting, as it minimizes the disruptive effects on the swiftlet population, whilst at the same time ensuring a high economic yield is delivered from the wildlife resource. Sodhi & Er (2000) recommended that governments, in partnership with industry, should work together towards the sustainable management of swiftlets in Southeast Asia. Better understanding of swiftlet taxonomy and ecology are needed to devise the most appropriate sustainable-harvest models. As with the bush meat crisis, social issues (e.g. alternative sources of income) may also need to be considered if large-scale sustainable harvesting of these imperiled species is to be achieved. Nevertheless, both the maleo and swiftlet examples show that the alleviation of overexploitation is possible.

How likely are the maleo and edible-nest swiftlets to be driven to extinction by human processes? This question can be answered, at least approximately, with the use of inferential population modelling. Recently one of the coauthors (Brook), derived a multivariate generalised linear mixed model that approximated statistically the estimated minimum viable population size (MVP) for 1198 well-studied species by relating extinction thresholds estimated through population viability analysis (PVA) to a suite of ecological 'correlates', including body size, generation length, niche breadth, reproductive rate, dispersal ability, range size and ecological flexibility

(B. W. Brook *et al.*, unpublished manuscript, 2004). An MVP can be broadly defined as the population size required for a species to have a predetermined probability of persistence over a certain length of time, given the inherent uncertainty in demographic, genetic and environmental processes that affect populations. Similar concepts are inherent in the IUCN categories of endangerment; e.g. following criterion E, a 'least concern' population has a greater than 90% probability of survival over a 100 year period (IUCN 2004). An inference based on the known life-history and environmental attributes, and the model of Brook *et al.* (unpublished manuscript) suggests that an MVP with a less than 10% probability of extinction over the next 100 years is at least 2300 individuals for the maleo, and 9100 for the swiftlets. These MVPs represent viable populations in the absence of external deterministic population pressures such as overexploitation by humans and habitat loss. Thus the maleo, with its current population size of 4000 to 7000 breeding pairs, is already worryingly close to its MVP, and will in all likelihood be driven to extinction within the next century if current anthropogenic pressures cannot be mitigated. The swiftlets, still numbering in the order of 50 million birds (Er *et al.* 1995), are not presently threatened with extinction, but may nevertheless suffer severe localised reductions in density that result in numerous local extinctions, which may eventually imperil the species.

Birds are not the only group being heavily hunted in Southeast Asia. In the Tangkoko division of the Tangkoko-Batuangas-DuaSudara Nature Reserve, endemic Sulawesi crested black macaques (*Macaca nigra*, a type of monkey) have declined substantially between 1978 and 1994 (Rosenbaum *et al.* 1998) (Fig. 4.7). One of the primary reasons for this decline appears to have been overhunting: these macaques have been found caught in snare traps set for forest (wild) pigs (*Sus scrofa*). There are predictions that this macaque will be wiped out within the next 20 years if the current levels of hunting continues unabated (Lee *et al.* 1999). In the same area, other researchers determined how hunting and habitat change were affecting the populations of a range of birds and mammals (O'Brien & Kinnaird 1996). In general, those species that were known to be hunted (e.g. Sulawesi crested black macaque) declined between 1979 and 1994, whereas those thought to be unaffected by hunting (e.g. Sulawesi taritic hornbill, *Penelopides exarhatus*) did not (Table 4.1).

The latter species are considered less vulnerable because they are not overly susceptible to the hunting techniques used predominantly in the study area (i.e. traps, snares, air-powered pellet rifles and dogs). The ground-nesting maleo, as described earlier, is a heavily exploited species on the verge of local extinction due to heavy egg collection, with only seven

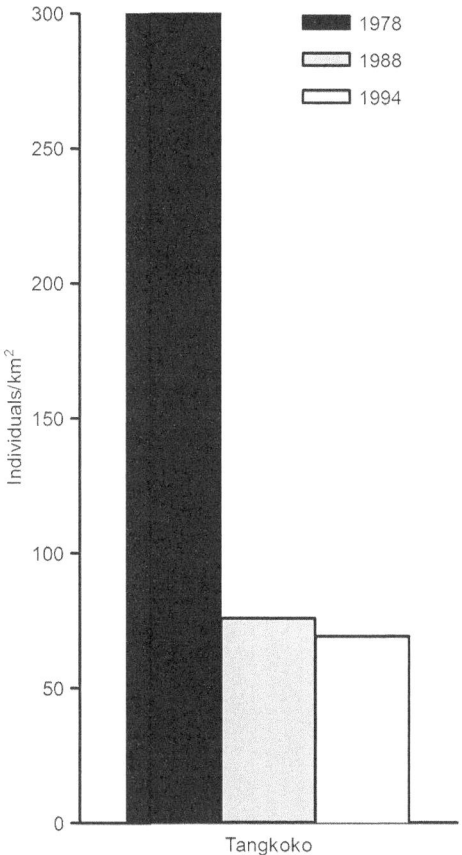

Figure 4.7 Decline in population density of the Sulawesi crested black macaques (*Macaca nigra*) between 1978 and 1994. (Modified from Rosenbaum *et al*. 1998. Copyright 1998 American Journal of Primatology. Reprinted with permission of Wiley-Liss, Inc., a subsidiary of John Wiley & Sons, Inc.)

pairs left. These observations suggest that excessive hunting has been responsible for population declines of some of the species in this area. O'Brien & Kinnaird (1996) recommended that urgent efforts are needed to curtail hunting to mitigate this loss, both in the study area and in the wider region.

Does hunting by traditional indigenous people substantially impinge on wildlife populations? Studies were carried out on the impact of traditional subsistence harvest by indigenous people called the Wana on birds in Sulawesi (Morowali Nature Reserve) (Alvard & Winarni 1999). There were several thousand Wana living in and around the reserve. These people

Table 4.1 *Changes in bird and animal populations due to hunting in Sulawesi*

Common name	Species	Expected response to hunting	Observed change
Anoa	*Bubalus depressicornis*	Decline	Decline
Sulawesi pig	*Sus celebensis*	No change/decline	No change
Crested black macaque	*Macaca nigra*	Decline	Decline
Bear cuscus	*Phalanger ursinus*	Decline	Decline
Babirusa	*Babyrousa babyrussa*	Decline	Decline
Maleo	*Macrocephalon maleo*	Decline	Decline
Tabon scrubfowl	*Megapodius cumingii*	No change	Increase
Red junglefowl	*Gallus gallus*	No change/decline	Decline
Red-knobbed hornbill	*Aceros cassidix*	Decline	Increase
Sulawesi taritic hornbill	*Penelopides exarhatus*	Decline	Increase

Reprinted from O'Brien & Kinnaird (1996).

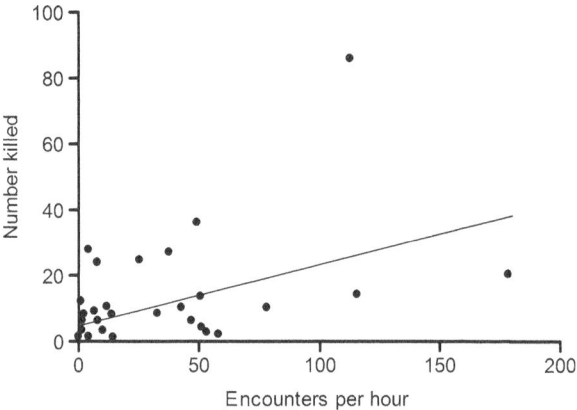

Figure 4.8 Relationship between species-abundance (as measured by the number of encounters per hour of transect) and the relative size of the harvest. Hunters tend to take more abundant species in greater number ($R^2 = 0.22$; $p < 0.003$) (Reprinted from Alvard & Winarni 1999). With permission from *Tropical Biodiversity*.

have very little contact with the outside world, practice slash and burn horticulture, and obtain their protein by hunting and trapping animals. Their animal prey includes mice, birds, bats, reptiles, amphibians, ungulates and primates. Blowguns are used primarily for hunting birds. Alvard & Winarni (1999) found that the Wana's common bird prey included flower-peckers, sunbirds, white-eyes and parrots. The more abundant a bird species was, the higher were its chances of being hunted by the Wana (Fig. 4.8).

Hunting does not seem to seriously affect the viability of bird populations in this area. Of the 46 prey species, 31 were more abundant at Wana
occupied sites than in regions not experiencing bird hunting. Although
habitat heterogeneity and differential detectability can also influence such
a result, this study does suggest that traditional hunting practices may not
cause heavy declines in prey (bird) populations. In another, similar finding
to the above, a low density of Aboriginal people using primitive hunting
techniques (e.g. snares) with only limited access, appeared to be having little
impact on tigers and their prey in Peninsular Malaysia (Tamah Negara
National Park) (Kawanishi & Sunquist 2004).

The above cases may nevertheless be anomalies rather than the norm for the
region. For instance, a group of indigenous Penan people appear to have been
responsible for the disappearance of an entire population of Bornean gibbons
(*Hylobates muelleri*) from a primary forest in Sarawak (Bennett *et al.* 2000).
Other studies report similar predicaments. In addition to habitat loss, the main
threat to the mammals of Sangihe and Talaud Islands (Indonesia) has been
hunting. Of 57 farmers interviewed, 77% admitted to hunting wildlife (Riley
2002). Hunting was the major threat to at least two mammalian groups: fruit
bats and cuscus (Fig. 4.9). Cuscus, an arboreal possum-like animal, has a
number of sub-species endemic to this region (e.g. bear cuscus, *Ailurops ursinus
melanotis*) and 5 of the 11 fruit bats are listed as globally threatened (e.g.
Talaud Islands flying fox, *Acerodon humilis*) (Riley 2002). Farmers predominantly used mist-nets and snares to catch prey (Fig. 4.9). The bats so harvested
are regularly sold in some markets in Indonesia (Fig. 4.10).

Riley recommended that on the islands, efforts should be made to better
regulate hunting and minimize the further loss and degradation of local
forests. Increase in community awareness concerning the plight of wildlife
and stricter law enforcement were required. However, as discussed earlier,
complicated social issues (e.g. finding alternative protein source) need to be
taken into account when attempting to achieve tangible conservation.

Up to 90 000 mammals are sold annually in a single market in north
Sulawesi (Clayton & Milner-Gulland 2000). It has been estimated that
86 198 kg of wild prey was consumed by people living around Manembonembo
and Gunung Ambang Nature Reserves in Sulawesi (Lee 2000). In 9 markets
in north Sulawesi, 24 864 individuals of 27 mammal species were sold (Lee
et al. 1999) (Fig. 4.11). Of these, approximately 1% of individuals harvested
were from protected species (e.g. dwarf cuscus). This apparently low proportion still represents many hundreds of individual animals, which for species
with small, already threatened populations, may be sufficient to eventually
lead to their extinction (Lee 2000). Creation of highways in Sulawesi facilitates this wild meat trade by allowing quick transport to the markets.

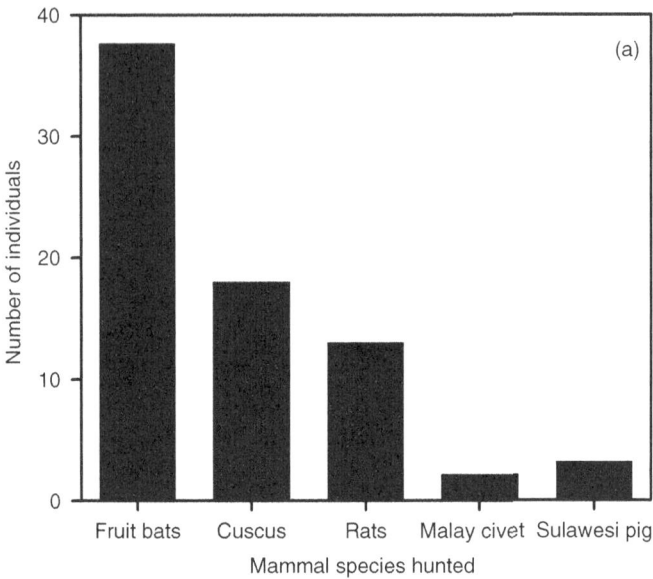

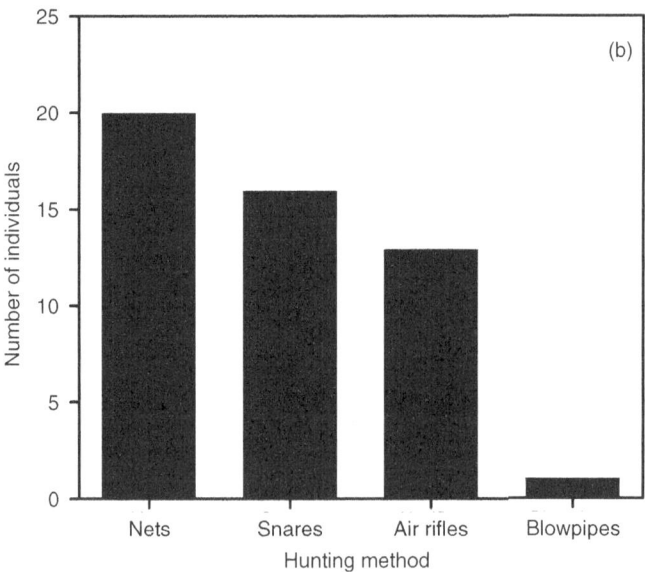

Figure 4.9 (a) Number of individuals of different mammal species hunted.
(b) Number of individuals hunted by different hunting methods. (Data from
Riley 2002.)

Figure 4.10 Harvested bats being sold in markets in Sulawesi.

A spatial model has been developed to describe the hunting of two endemic pig species, the babirusa (*Babyrousa babyrussa*) and the Sulawesi wild pig (*Sus celebensis*) in Sulawesi (Clayton *et al.* 1997). They found that hunting will more heavily impact the former species than the latter, because babirusa are restricted only to the rain forest, are suffering from rapid range contraction due to loss of habitat, and have a total wild population estimated to be only 5000 individuals (Clayton *et al.* 1997) The Sulawesi wild pig is not protected by law, but the babirusa is. The penalty for killing a babirusa is considerable, with a maximum fine of US$45 454 or 5 years in prison. However, the chances of a dealer actually being fined are minimal, because they routinely bribe the relevant officials with meat (Clayton *et al.* 1997). Clayton *et al.* (1997) recommended that the most realistic means of curtailing the hunting of babirusa will be to religiously impose fines on illegal traders attempting to sell babirusa meat at the markets. They also predicted that future increases in economic conditions may eventually lead to a decrease in the hunting pressure; a hypothesis not supported by a study from Borneo showing that even affluent people continued to hunt wildlife (Bennett *et al.* 2000).

Overexploitation of animal populations in tropical forests is not driven by a need for bush meat. For instance, over 500 000 wild reticulated pythons

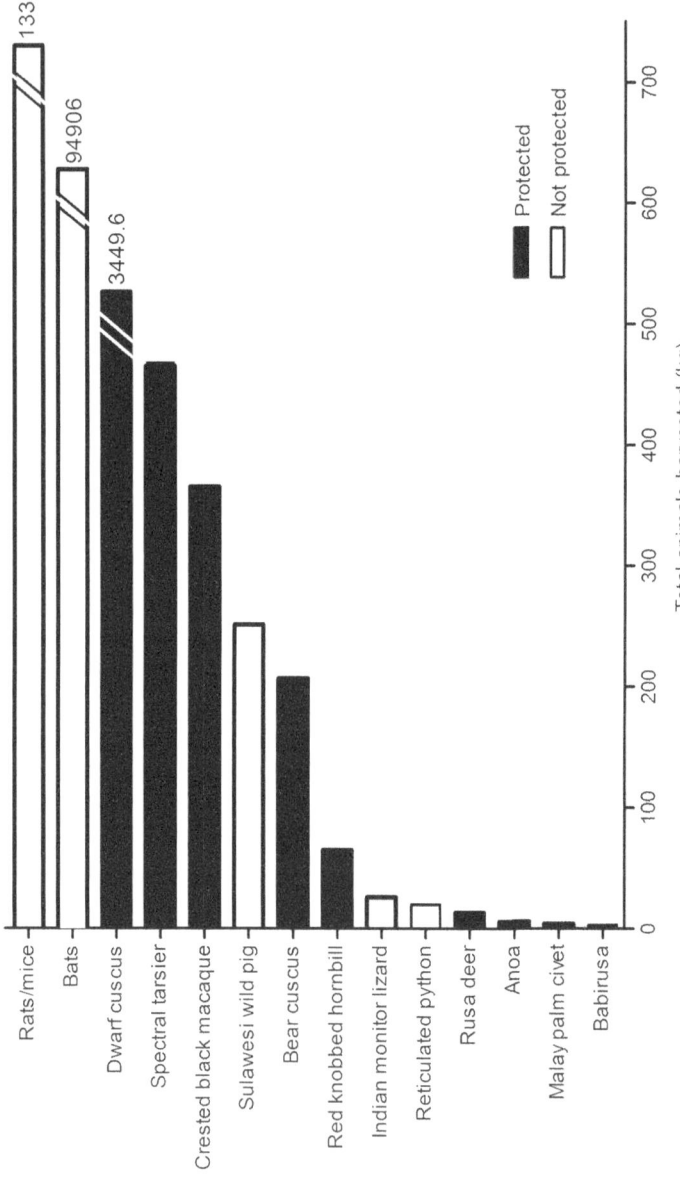

Figure 4.11 Harvested animals being sold in nine markets in north Sulawesi. (Data from Lee *et al.* 1999.)

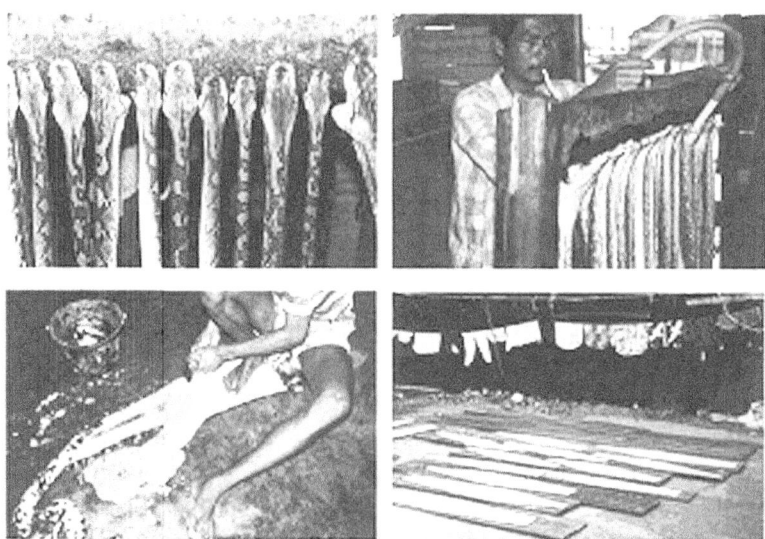

Figure 4.12 Harvest of wild reticulated pythons (Reprinted from Shine *et al.* 1999). With permission from Elsevier.

Figure 4.13 A python being sold by the highway in the Philippines. Pythons are regularly sold there to keep as pets or for meat.

(*Python reticulatus*) are harvested each year for the leather industry, which places a high value on their attractive skins for luxury goods (Groombridge & Luxmoore 1991) (Figs. 4.12 and 4.13). Information has been collected on the sizes, sexes, reproductive status and diet of 784 slaughtered pythons in

Sumatra (Shine *et al*. 1999). They concluded that because of intrinsic features of their biology, such as rapid growth, early maturation, high fecundity and ability to evade detection, even quite high levels of hunting are unlikely to cause their extirpation from Sumatra. However, since pythons also act to keep rodent populations in check, harvests which lead to a decline in python densities may have the unwanted side-effect of increasing concomitantly the damage to crops by rodents (Shine *et al*. 1999).

Trade for cage and medicinal use

Southeast Asia is a major hub of the global trade in wildlife and wildlife products, functioning as both consumer and supplier (WCMC 2002; Table 4.2). Worldwide, between 1.6 and 3.2 million birds are taken annually from wild populations for the pet trade (Beissinger 2001). Most (95%) of these birds represent finches and parrots. A rampant trade in live birds operates in Southeast Asia, affecting many rare or little known species, and even those protected by national laws (Corrigan 1992). Within the region, Indonesia represents a 'hot spot' of bird capture. From 1981 to 1992, roughly 100 000 birds were reported to be exported from Indonesia, but this number is, in all likelihood, a gross underestimation of the actual numbers of wild birds caught (Director General of Forest Protection and Nature Conservation [PHPA] 1998). Many birds are likely to remain uncounted (e.g. due to illegal capture) or have died before reaching the point of official tallying, or else are sold in local markets. It is estimated that up to 60% of birds trapped may actually die before reaching the export market (Inigo-Elias & Ramos 1991). As with the bush meat issue, the biological understanding required to guide the sustainable extraction of wild birds is generally lacking (Beissinger 2001).

The critically endangered Bali starling (*Leucopsar rothschildi*) epitomises how excessive capture for the pet trade led to a species' endangerment. Currently, there are only six wild individuals of this species left, restricted to the Bali Barat National Park (van Balen *et al*. 2000); http://www.birdlife. org). Rampant trapping coupled with habitat loss resulted in the precipitous population decline of the Bali starling. This species was listed on Appendix I of CITES in 1970, and subsequently received official protection in Indonesia in 1971. Despite legal protection, 19 individuals were witnessed being sold illegally in shops in Singapore in 1979, and 16 individuals were observed in cages in Denpasar (Bali) in 1982 (van Balen *et al*. 2000). Extremely small population size, distribution limited to one site, a continued diminishment of its natural habitat, and possible ongoing illegal trapping have all conspired to drive this species to the brink of extinction. If this species is to have any realistic chance of long-term persistence, efforts such as 'real'

Table 4.2 *Net legal export in selected wildlife products*

Country	Live lizards	Live snakes	Live primates	Live parrots	Lizard skins	Snake skins	Crocodilian skins	Cat skins
Myanmar	—	—	4 (<0.1)	−67	—	—	—	—
Laos	—	6000 (2.4)	—	—	2 (<0.1)	2 (<0.1)	—	—
Vietnam	−18	35484 (14.2)	3149 (8.9)	2751 (0.5)	—	109426 (8.4)	—	32 (<0.1)
Thailand	−3	−21	−63	−2587	—	42533 (3.3)	−5595	—
Cambodia	—	—	200 (0.6)	—	—	—	—	—
Malaysia	949 (0.1)	15713 (6.3)	−76	−11297	252253 (16.2)	520776 (40.1)	−312	—
Singapore	−52	−1474	−83	−5484	60843 (3.9)	202857 (15.6)	38208 (4.9)	−3
Indonesia	7474 (0.8)	22399 (9.0)	3324 (9.4)	−25025	457600 (29.4)	366118 (28.2)	10385 (1.3)	—
Brunei	—	—	—	−500	—	—	—	—
Philippines	−510	−60	2085 (5.9)	−788	−825	1	−185	—

Negative numbers represent net number of items imported; numbers in parentheses represent percentage of the respective items exported in the world (WCMC 2002; Reprinted from Sodhi *et al.* 2004b. With permission from Elsevier).

(enforced) protection, habitat preservation and enhancement through reve-getation, and possibly a release of captive bred individuals to bolster wild numbers, are urgently needed.

As with the Bali starling, capture of the yellow-crested cockatoo (*Cacatua sulphurea*) is the main factor causing its decline in Indonesia (Director General of Forest Protection and Nature Conservation [PHPA] 1998). The recovery of this species can be brought about by curtailing or halting illegal capture, and through the provision of adequate habitat (e.g. cavity producing trees for nesting). It is rather ironic that accidental or deliberate releases of yellow-crested cockatoos have led to its establish-ment in new areas outside of its native range, such as Singapore and Hong Kong. There has been concern that in these new areas, this species can outcompete and displace native birds (Sodhi & Sharp 2005).

Notwithstanding this apprehension, the establishment of yellow-crested cockatoos in new areas does show that captive release can be a viable strategy to reinforce wild populations, provided that wild capture is pre-vented and adequate habitat is available.

As is the case in Indonesia, native birds, macaques and gibbons (espe-cially the lar gibbon, *Hylobates lar*) are commonly kept as pets in both rural and urban areas of Thailand. Forest loss has made gibbons particularly vulnerable to trapping (Eudey 1994). Abandoned pets often end up in temples or government agencies. Possible reintroduction of rehabilitated gibbons can be envisioned to boost their wild populations (Eudey 1994).

Not all capture of wildlife is for the pet trade. Many animal and plant products form components of traditional Chinese/Asian medicine, which dates back more than 5000 years. Harvest for Chinese medicine may already be responsible for the depletion of wild populations of species such as tigers, bears, rhinos and swiftlets. This off-take is exemplified by the Sumatran tiger (*Panthera tigris sumatrae*), whose body parts (e.g. bones and penis) are used in traditional Asian medicine. The continued medicine-related demand for this species is very worrying, as it is now listed by the IUCN as critically endangered, with fewer than 500 individuals thought to be left in the wild (Seidensticker *et al.* 1999). Despite Sumatran tigers enjoying full legal protection, with penalties of jail times and heavy fines, a survey in 2002 revealed a continuing substantial commerce in tiger parts and products in Sumatra (Shepherd & Magnus 2004). Of 24 towns and cities surveyed, 17 (71%) had shops selling tiger parts, with one-quarter of 117 shops and dealers located therein having tiger parts for sale. Most tigers were killed by professional or semi-professional hunters using inexpensive and simple-to-use wire cable log snares. Although it could be argued that at least some tigers were killed legitimately because they attacked humans and livestock, most were

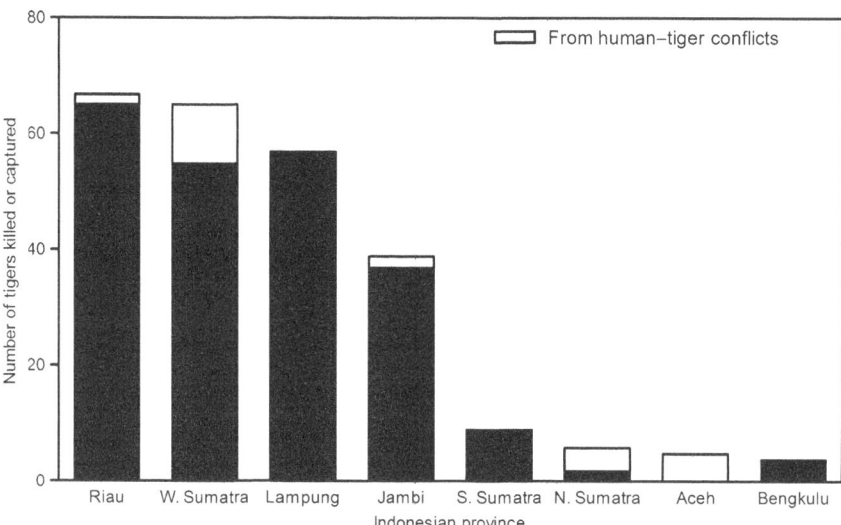

Figure 4.14 The number of tigers killed or captured in Indonesia between 1997 and 2002. (Data from Shepherd & Magnus 2004.)

taken for commercial purposes (78% of 51 estimated tigers killed per year in Sumatra) (Fig. 4.14). Because there is no programme in place to compensate local people for loss of livestock or life due to tiger attack, the illegal sale of tigers killed after they have come into conflict with humans remains the only way for people to recuperate their losses (Shepherd & Magnus 2004).

Over the 18 years between 1975 and 1992, South Korea imported an astounding 6128 kg of tiger bones; an average of 340 kg per year (Mills 1993). The greater majority (61%) of these imported bones originated from Indonesia. Economics is postulated to govern hunting effort (Robinson 2001). Intuitively, the scarcer a species is, the more expensive it becomes, thus enhancing the incentives for poaching and creating a vicious feedback loop that may ultimately end in the harvested species' demise. This worrying scenario indeed seems to be the case for tiger bone exports: there was a strong negative correlation between the price of bones imported into South Korea from Indonesia and their total weight (Fig. 4.15). Shepherd & Magnus (2004) recommended that better protection of wildlife reserves for tigers, public education, proper law enforcement and addressing of social issues (e.g. alternative employment means for traditional people) hold the only hope of curtailing tiger hunting in Sumatra and thereby helping recover this magnificent animal. It is sad to see that tigers are still being hunted illegally for meat in Peninsular Malaysia mainly to cater for the palates of rich Singaporeans (Anonymous 2004c).

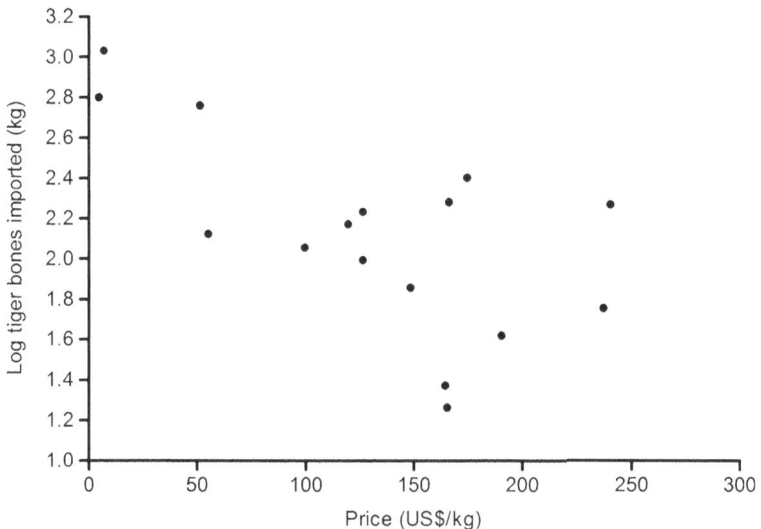

Figure 4.15 Correlation between price and weight of tiger bones imported into South Korea from Indonesia ($r = -0.65$; $p = 0.007$; $n = 16$). (Data from Shepherd & Magnus 2004.)

Like tiger body parts, rhino horn has been used in Chinese medicine since at least 2600 BC, and has led to the near extinction of the Javan (*Rhinoceros sondaicus*) and Sumatran rhinos (*Dicerorhinus sumatrensis*). Heavy trading in rhino horns occurred during the T'ang Dynasty (Schafer 1963; Nowell *et al.* 1992). Poaching, coupled with habitat loss, has driven rhinos to the brink of extinction in Southeast Asia. Although protected by law in both Indonesia and Malaysia, as with other cases mentioned above, this legislation is rarely effective in curbing the poaching of Sumatran rhinos (Rabinowitz 1995). Inadequate monitoring and subsequent protection in key areas, and futile efforts at captive breeding, all failed to halt the decline of this species. The ongoing inability of the responsible national governments and international funding agencies to take tangible steps are likely to make extinction a reality for this species (Rabinowitz 1995).

Farming of species used in traditional medicine, or finding alternatives to the use of animal parts of threatened species, are both possible alternatives to hunting that may reduce the pressure on precarious wild populations (Mainka & Mills 1995; von Hippel & von Hippel 2002). However, if farming is undertaken, efforts must be put in place to monitor animal welfare. Further, public education and proper wildlife protection schemes are also required for the effective alleviation of poaching pressure for medicinal use on the threatened wildlife species of the region.

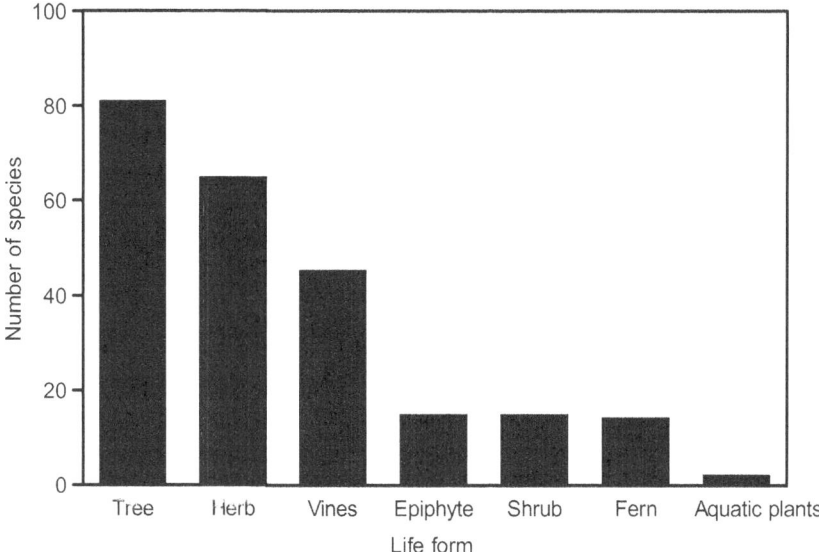

Figure 4.16 The number of species of plant life-forms used for medicine in a study site in Kalimantan, Borneo. (Data from Caniago & Siebert 1998.)

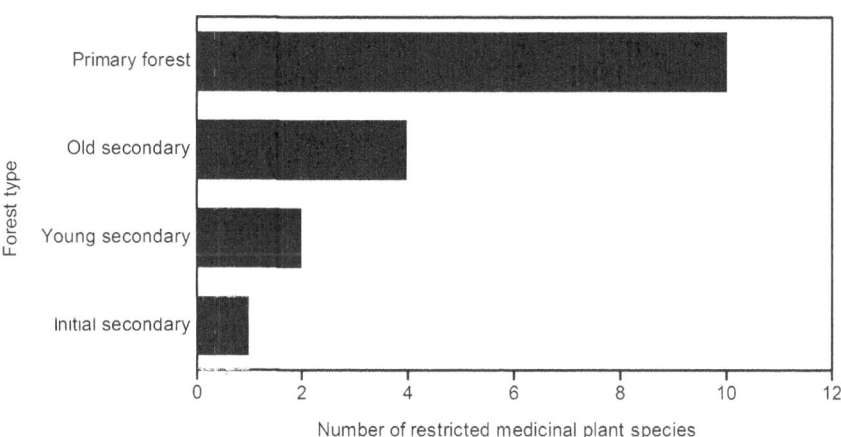

Figure 4.17 Number of medicinal plant species restricted to different forest types. (Data from Caniago & Siebert 1998.)

In addition to animals, medicinal plants are regularly used by various locals in Southeast Asia. In Kalimantan (Ransa Dayak village), 237 medicinal plants were recorded as being exploited for such purposes, of which tree species were the most frequently (81) used (Fig. 4.16) (Caniago & Siebert 1998). Compared to other forest types, primary forests had the highest number of exclusive medicinal species (Fig. 4.17). Since there may be few alternative species to medicinal plants of the primary forest, the loss of this habitat type through logging and other human activities (see Chapter 1) rather paradoxically poses a grave challenge to their ongoing use for traditional medicine (Caniago & Siebert 1998).

Invasive species

The negative impacts of biological invasions on native biodiversity are perhaps as severe as other high profile threats such as global climate change and habitat clearance (Vitousek *et al.* 1996). Modern trade and transport has facilitated the spread of biological invaders. It is estimated that current annual damage caused by biological invaders in the United States alone is US\$137 billion (Pimental *et al.* 2000). This figure includes economic losses and control costs, but does not take into account the unmeasurable costs of biodiversity losses (e.g. loss of ecosystem services). With increasing commerce and habitat loss, invasive species (also broadly called exotics or introduced species) may pose a serious potential threat to the future of tropical biodiversity. However, relatively little is known on the extent of damage caused by invasive species on Southeast Asian biotas.

Invasive plants can be a major competitor with native vegetation, and in extreme cases, can convert large areas to exotic monocultures. Tropical rain forests are the most speciose of all terrestrial ecosystems, and it is traditionally believed that their high species-richness provides a good barrier to biological invasions (Pimm *et al.* 1991). Therefore, the number of invasive species in a tropical forest can be a good indication of its level of deterioration. In Singapore, there were no introduced species found in mangroves and only a single tropical American plant species – a bird dispersed shrub called Koster's curse (*Clidemia hirta*) – in primary and tall (15–25 m) secondary forests (Teo *et al.* 2003). This is surprising, because Singapore is heavily deforested with a matrix (areas surrounding forest patches) rich in cultivated and naturalised exotic species (Corlett 1988). The study by Teo *et al.* (2003) supports the hypothesis of the likely lower invasibility of Southeast Asian rain forests. One of the factors governing this low invasibility is probably the presence of closed canopy, because the number of exotic species is reported to increase as the canopy becomes more open (Fig. 4.18) (Teo *et al.* 2003).

As with plants, it was found that those forest patches in Singapore with a closed canopy (primary and late secondary forests) supported fewer introduced bird species than abandoned open-canopy plantations and young secondary forests (Castelletta *et al.* 2005). Similarly, over a 100 year period (1898–1998), a small 4 ha patch of residual forest in the Singapore Botanic

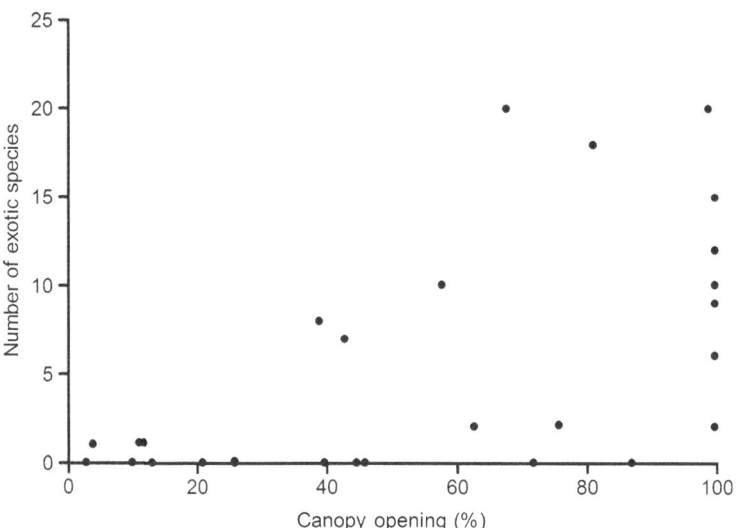

Figure 4.18 Relationship between the number of exotic plant species and canopy opening. (Reprinted with permission from Teo *et al.* 2003.)

Gardens became home to significantly more introduced species (e.g. the house crow, *Corvus splendens*) (Sodhi *et al.* 2005a) than other larger areas of primary forest. In 1998, 20% of bird individuals in this patch were exotics, suggesting a degradation in habitat quality over time, and a greater exposure to invasions due to more pronounced edge effects. Sodhi *et al.* (2005a) hypothesised that high occurrence of introduced birds in this fragment may eventually spell doom (e.g. through higher nest predation) for its existing native birds.

As with the forest plants and birds, forest fishes also seem to be largely immune to competition with invasive species in Singapore. A total of 37 invasive species of freshwater fishes, turtles, molluscs and prawns have now been established on the island (Ng *et al.* 1993), yet these exotics do not appear to have affected the biodiversity of the aquatic fauna of forested streams. One of the reasons may be that a large proportion of the native freshwater fauna of Singapore (i.e. 80% of 54 freshwater fishes) prefer acidic waters, whereas introduced aquatic fauna are largely restricted to neutral or alkaline waters. However, a recent invasion of species that can thrive in acidic waters (e.g. tiger barb, *Puntius tetrazona*) is a cause of concern. Therefore, Ng *et al.* (1993) recommended that tightly enforced legislation and strict quarantine measures are needed to ensure that no other acidic water preferring species become established in Singapore, and attempts should be made to eradicate existing acid water exotics.

As in Singapore, the Koster's curse was found in an undisturbed forest in Peninsular Malaysia (Pasoh Forest Reserve), being first discovered there in the early 1990s (Peters 2001). A survey in 1997 of a 50 ha plot in Pasoh recorded 1002 individuals, of which 7% were in reproductive condition. All but eight of these plants were recorded in forest gaps and gap edges. The location of Koster's curse in gaps was facilitated by disturbance caused by wild pigs, an animal now super abundant in the Reserve (see Chapter 3). No reproductive individual of Koster's curse was found in the forest under-storey. Peters (2001) postulated that Koster's curse will outcompete the native plants in forest gaps and thus can alter the process of forest regeneration, following events such as natural treefalls or selective logging. This study shows that alteration of biotic interactions (e.g. super abundance of pigs) in a forest can have unanticipated knock-on effects (i.e. changes in plant community composition). However, it would be interesting to carry out further studies to monitor the spread of Koster's curse at this site, to see whether the competitive take-over prediction of Peters (2001) actually eventuates.

One can assume that if the same invasive species is shown to be creating havoc in a number of different areas, it is also likely to be having, or will have, an impact throughout Southeast Asia. As mentioned, invasive ecology is still in its infancy in this region. Below we review some of the likely detriments of invasive birds to native birds, which is the best studied taxonomic group. The avifauna of Southeast Asia is currently threatened by habitat loss (Laurance 1999; Achard *et al.* 2002; Brook *et al.* 2003). The presence of, and potential increase in populations of invasive birds could compound the survival pressures on native avifauna through predation, disturbance or competition for resources.

The invasive common myna (*Acridotheres tristis*) nests in tree hollows and may compete for these with native hollow-nesting species (Pell & Tidemann 1997). It has been hypothesised that one of the factors respon-sible for the decline of the hole-nesting oriental magpie robin (*Copsychus saularis*) in Singapore might be the spread of mynas (Huong & Sodhi 1997). The nesting of the common myna on the same trees was also known to disturb the nesting of the threatened Seychelles magpie robin (*Copsychus sechellarum*) on Fregate Island in the Seychelles. These nest disturbances had an adverse effect on the breeding success of the robins (Komdeur 1996). Therefore, common mynas, now flourishing in many open or urban habitats of the region, can possibly pose a threat to native cavity-nesting birds. In Kenya, the house crow was observed to raid the nests of ploceid weavers and other small bird species (Ryall 1994), and could pose a similar threat to native parkland birds (e.g. yellow-vented bulbul, *Pycnonotus goiavier*) in

Southeast Asia. In addition to predation and competition, some invasive species (e.g. rock pigeon, *Columba livia*) can hybridise with wild pigeons (Johnston & Janiga 1995), causing genetic dilution and potentially leading to the loss of locally evolved adaptations.

Climate change

Earth's climate has warmed by about 0.6 °C during the past century (Zwiers 2002). This change in global climate is now recognised as one of the most important factors in altering the distribution and abundance of biotas across the planet (Warren *et al.* 2001). Species have been reported to change their elevational distribution and latitudinal ranges towards cooler locations, likely in response to global warming (Grabherr *et al.* 1994; Parmesan *et al.* 1999). Southeast Asian vegetation probably fluctuated considerably (i.e. the extent of rain forest) over the last 65 million years due to periodically changing climate (Heaney 1991). However, the extent to which the latest and most rapid phase of global warming is impacting the biodiversity of Southeast Asia is poorly known. By comparing data from two field guides of Southeast Asian birds that were published in 1975 and 2000, K. S. Peh & N. S. Sodhi (unpublished data 2005) found that elevational changes in distribution may have occurred in 138 resident bird species. Changes in the elevation were caused by shifts in either (or both) of the upper and lower altitudinal boundaries. Eight species exhibited a marked response, by shifting upwards both their upper and lower altitudinal boundaries (e.g. the little forktail, *Enicurus scouleri*). Peh & Sodhi argued that a large proportion of species were likely to have shifted their ranges towards a higher elevation in response to climate warming, though other explanations are admittedly possible. The fact that these elevation-related changes occurred even in habitat generalists reinforces the notion that habitat loss may not be the sole reason for the elevation shifts, because these species are relatively insensitive to changing land use. Similarly disturbing shifts in the altitudinal range of amphibian species have also been recorded in the rain forests of tropical north Queensland, Australia (Thomas *et al.* 2004). Southeast Asia has warmed by at least 0.3 °C in the past two decades and temperatures are projected to increase by 1.1–4.5 °C by year 2070 (IPCC 2001). Peh & Sodhi's study implies that climatic warming may very well turn out to be an important factor affecting the already imperiled biotas of Southeast Asia, especially in montane areas.

In addition to climate warming, other atmospheric changes such as elevated atmospheric CO_2 and increases in nitrogen deposition can also affect biodiversity negatively and disrupt ecological processes. In other tropical and temperate biomes, elevated CO_2 levels have been demonstrated to, or are predicted to, reduce species diversity, alter biotic interactions and facilitate the spread of invasive species, especially woody weeds (Coley 1998; Smith *et al.* 2000; Hartley & Jones 2003; Zavaleta *et al.* 2003). It has also been suggested that the negative effects of the elevation in CO_2 levels may be higher in more speciose communities, as these may contain more responsive species than species-poor communities (Niklaus *et al.* 2001). However, admittedly little is known with any degree of certainty regarding how climatic and atmospheric changes will impact Southeast Asian biotas. Clearly, more studies are needed in this direction.

Zoonotic disease threats and likely associated slaughter of wildlife

In the past 30 years, a large proportion of diseases have apparently entered the human population through contact with wildlife. Roughly two-thirds of human pathogens are zoonotic, meaning that they are able to pass from animals to humans, or vice versa. Southeast Asia, has recently witnessed three such relatively major pandemics: Nipah virus, SARS (Severe Acute Respiratory Syndrome) and bird flu. The probable role of animals in the spread of all three diseases (see beyond), and subsequent large scale culling that can result from such outbreaks, can exert a pressure on already imperiled populations. Below we briefly highlight some of the cases for concern.

Between September 1998 and 1999, a disease emerged in Peninsular Malaysia that resulted in the death of 105 humans and the slaughter of over 1 million domestic pigs (*Sus scrofa*) (Yob *et al.* 2001). The culprit in this case was discovered to be a new disease, subsequently named as Nipah virus, named after the town where it first appeared (Chua *et al.* 2000). The presence of Nipah virus was subsequently discovered in four species of fruit bats (Megachiroptera) collected widely from Peninsular Malaysia (Yob *et al.* 2001). The infection rates were 4 to 31% for different species. In addition, an individual of an insectivorous bat species, the lesser Asian house bat (*Scotophilus kuhlii*), was also infected. It is hypothesised that the smoke from Southeast Asia's massive forest fires in 1997–8 (see above) probably caused the phenological failure of many trees and a resultant low production of forest fruits. To find food, many fruit bats switched to the fruit trees in Malaysia's large pig farms. These bats were likely to have passed on the deadly Nipah virus to the pigs, which then transmitted it as an intermediary to humans. That transmission forced the government to destroy huge numbers of pigs (Chivian 2002), and cost millions of dollars in lost production. Although there is no unequivocal evidence that bats can spread Nipah virus directly to humans, the mere presence of this virus in them raises a concern that should there be another epidemic, there could be calls to (indiscriminately) cull bats. Considering Peninsular Malaysia harbours up to seven threatened bat species (IUCN 2003), were it to happen, it would be likely to devastate bat populations already facing a precarious situation due to massive habitat loss.

SARS is another case for concern from the biodiversity angle. With over 8000 people infected and about 774 fatalities, SARS had been a formidable hazard to human health (http://www.who.int/csr/sars/country/table2004_04_21/en/). This disease first originated in Guangdong Province, China, in late 2002, and quickly spread to at least 30 countries by 2003. Except Toronto (Canada), all SARS hotspots were in Asia (China, Hong Kong, Taiwan and Singapore). Caused by a coronavirus related to influenza, the origin of SARS remains unclear, but an animal origin is suspected. The recent discovery of a SARS coronavirus in wild Chinese animals like the masked palm civet (*Paguma larvata*) and raccoon dog (*Nyctereuteus procyonoides*) implicates them in the outbreak (Guan *et al.* 2003). The animals tested were caught from the wild and were part of a booming market for wildlife meat in East Asia. Although it remains possible that these sampled individuals were infected by humans, reports such as these may nevertheless have dire consequences for small carnivores in Asia. Many small Asian carnivores are endangered (e.g. 20% of civets are threatened by extinction) (IUCN 2003) and the more common ones play a critical role in forest ecosystems, as they keep the population of other small mammals, including rodent pests, in check. Their demise or population depletion could theoretically result in an explosion of small mammal populations, which could, in turn, prey upon bird eggs and seeds. The potential implications of such knock-on effects for both native biodiversity and human wellbeing are profound (Wong *et al.* 1998).

While scientists and the World Health Organization (WHO) maintain that the evidence thus far linking the spread of SARS with wild animals is weak and circumstantial, there is concern that certain sections of the press and politicians, searching for a scapegoat to prevent future SARS outbreaks, may misuse such information to argue for the mass slaughter of certain wildlife. We have earlier seen SARS-related panic culling of animals in China (http://news.nationalgeographic.com/news/2004/01/0109_040109_SARS.html). Similarly, a mass culling of stray cats (*Felis catus*), at least indirectly SARS-related, was envisioned for Singapore. Statements by disease experts in the Singapore press such as '... short of killing all the animals, there is no other way to do it' are concerning (Sim & Soares 2003). Some governments may heed this call by culling wild animals, in the interests of appearing to be proactive (in the public eye) about preventing the further spread of SARS. We urge agencies (e.g. WHO) to do more to educate governments and the public to avoid the rash and wholesale slaughter of already threatened wildlife of Southeast Asia that would otherwise take place should another SARS or SARS-like epidemic emerge in the region.

Late 2003 through to early 2004 saw witness to yet another zoonotic pandemic in Southeast Asia. Of 42 humans infected by bird flu in Thailand and Vietnam, 33 died (http://www.who.int/csr/disease/avian_influenza/ country/cases_table_2004_09_28/en/). Bird flu (H5N1) is a type of influenza that infects birds but can occasionally be transmitted from live birds to people. Human to human transfer of this disease is likely, but not proven conclusively (Ferguson *et al.* 2004). This disease resulted in the culling of over 100 million poultry in Asia. Concerns have been raised in the media that migratory birds can transmit the bird flu virus over long distances and excrete it in their droppings, thus facilitating its spread. According to WHO, migratory waterfowl are the natural reservoir for bird flu viruses, and these birds are also known to be resistant to the virus, developing only a mild and short-lived illness themselves (http://www.who.int/csr/disease/avian_in-fluenza/avian_faqs/en/). There are reports of the culling of wild birds in certain areas to be envisioned as a pre-emptive measure designed to halt the spread of bird flu (http://www.birdlife.org). There has been recent discovery of the bird flu virus in common urban/parkland resident bird species (e.g. rock pigeon, *Columba livia*) but not in migratory bird species (Ghosh 2004). Should another outbreak of bird flu occur in Southeast Asia during the migratory period of birds (September to April), with some migratory birds found to be carrying the virus, then the possibility of indiscriminate culling of migratory birds looms large. As with SARS and their potential mammalian hosts, this could be devastating for these birds' populations, considering that they are already threatened in Asia due to the chronic loss of wetlands and overharvesting (Asia–Pacific Migratory Waterbird Conservation Committee 2001). Again, agencies such as WHO and FAO should provide proper guidance that also takes into account animal welfare.

Whilst we do not wish to sound alarmist, we feel it is imperative to show how pandemics of zoonotic diseases may result in the mass slaughter of native animals, especially in poorer Southeast Asian countries where the population is expanding, public education levels are generally low, and world-class health-care is not widely accessible.

In addition to the transfer of diseases, direct human–animal conflicts can also be a hurdle to wildlife conservation. This is exemplified by crop raiding elephants (*Elephas maximus*) in Sumatra. There are between 2800 and 4800 wild elephants left in Sumatra. A study has been carried out on elephant–human conflict in the Way Kambas National Park, where 250 to 350 wild elephant exist (Nyhus *et al*. 2000). In addition, there were about 150 captive elephants in the Elephant Training Centre located in the park. Around the park, wild elephants damaged 45 ha of crops and 900 plantations (e.g. banana, *Musa* spp.) over an 18 month period. Further, one person was killed and another injured by elephants in this area. This conflict with elephants diminishes local support for the haven for wild elephants that the park currently provides. Nyhus *et al*. (2000) suggest that the risk of human–elephant conflicts needs to be minimised if the long-term viability of this park is to be guaranteed. They suggest constructing elephant exclusion trenches near 'conflict hotspots'; erecting electric fences, if necessary; and the provision of adequate compensation to the villagers for the damage caused by elephants.

Summary

1. Habitat loss and degradation are not the only threats facing Southeast Asian biodiversity. Forest fires, overexploitation for bush meat and the pet trade, invasive species, climate change and the likely animal slaughter to prevent the spread of zoonotic diseases, all represent serious threats, both singly and synergistically.

2. Increasingly severe El Niño events, coupled with poor land use practices, are driving an increase in the frequency of catastrophic forest fires in Southeast Asia. Fires, especially their repeated occurrence in rain forests, can diminish the chances of vegetation recovery and reduce the productivity of habitats on which the fauna depends.

3. Wildlife hunting for sustenance and cash are likely to be depressing populations of some charismatic Southeast Asian animals such as the Sumatran tiger and rhino, as well as many lesser known species. Deeper biological as well as sociological understanding should be used to find solutions to this dangerous predicament.

4. Capture of wildlife for the pet trade and medicinal uses has endangered some Southeast Asian animals and also most likely, plants.

5. Southeast Asian highly speciose biotas may be resistant, to some degree, to the effects of invasive species. However, this may change as forests are converted to open and degraded areas, and the buffers provided by a rich biodiversity are lost. More research is badly needed in this direction.

6. Although some Southeast Asian bird species appear to be shifting their distributional ranges to cooler climates in response to global

warming, more definitive and comprehensive studies of these impacts are required.

7. Recent pandemics of zoonotic diseases in Southeast Asia pose yet another threat to certain imperiled Southeast Asian animals via pre-emptive or post-hoc culling of wildlife populations. We suggest that agencies such as WHO should better educate the governments and people of the region to help alleviate this threat.

Chapter 5

The projected future of biodiversity in Southeast Asia

In this chapter we explore the future of Southeast Asian biodiversity. The rate and extent of human-mediated extinctions is intensely debated (Heywood *et al.* 1994; Dirzo & Raven 2003). Despite the debate, there is a general agreement that extinction rates have soared recently owing to accelerating habitat destruction and burgeoning human populations (Balmford 1996; Kerr & Burkey 2002). Humans may be directly or indirectly responsible for between 100 and 10 000 times more biotic extinctions than 'natural' or 'background' extinctions caused by factors such as speciation (Wilson 1989; Pimm *et al.* 1995; Dirzo & Raven 2003). Close to 600 species of threatened birds and mammals are predicted to become extinct over the next 50 years (Dirzo & Raven 2003).

Whether this extinction rate translates into actual species demises is again hotly debated (Heywood & Stuart 1992; http//www.pbs.org/wgbh/evolution/extinction/massext/index.html). Various estimates range between a few thousand to more than 100 000 species being extinguished every year (Ehrlich & Ehrlich 1996; Wilson 2000). Pimm & Raven (2000) estimated that currently 1000 to 7000 species are being lost per decade. They also predicted that the current rate of deforestation alone may wipe out 40% of the species in 25 biodiversity hotspots identified by Myers *et al.* (2000). Other studies predict likely extinctions in various taxonomic groups. For example, it is predicted that 350 species of birds (3.5% of the extant avifauna) may go extinct by 2050 (Jenkins 2003). Due to their requirements for a large area, the species at risk may be much higher for other taxonomic groups such as large mammals. Further, as reported in Chapter 3, within the same taxonomic group, a species, for example, because of its large size, may be more extinction-prone than others. All of the above estimates, we admit, may be prone to some error, but they do point towards a looming global biodiversity crisis (i.e. 'the sixth great extinction'; perhaps a more catastrophic one in the tropics where two-thirds of the biodiversity reside).

133

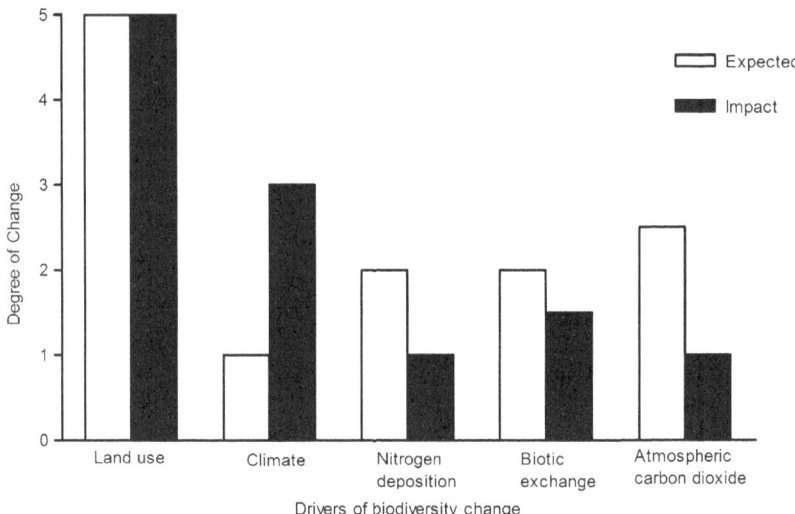

Figure 5.1 Expected changes for the year 2100 in the five major drivers of
biodiversity change and their likely impact on biodiversity in the tropics.
The ranks are given from 1 (low) to 5 (high) for each driver. (Data from
Sala *et al.* 2000.)

Despite being simplistic, in the absence of a better option, the species-area
equation is widely used to estimate species extinctions in relation to habitat loss.
At the estimated loss of 0.8% of forest per year globally, using the species-area
equation, Hughes *et al.* (1997) predicted that between 0.1 and 0.3% of tropical
forest species could go extinct every year. In other words, about 14 000 to
40 000 species could vanish from tropical forests every year (Hughes *et al.*
1997). Many species have distinct populations and these are crucial in maintaining the genetic heterogeneity. Hughes *et al.* (1997) estimated that 16 million
populations could be lost every year from tropical forests as a result of deforestation. Clearly, these figures are frightening and more so because 19 out of
20 species in tropical forests may be unknown to science (Dirzo & Raven 2003).

Deforestation is not the only driver of biodiversity change (see Chapter 4).
Sala *et al.* (2000) modelled the sensitivity of biodiversity in various biomes
to major drivers influencing its change: changes in land-use (habitat loss
and degradation), climate change (global warming), elevation in CO_2 levels,
increase in nitrogen deposition and biotic exchange (deliberate or accidental
introduction of non-native biotas). Sala *et al.* (2000) estimated that there
will be major changes in land-use in the tropics by the year 2100 and this will
exert a major negative impact on its biodiversity (Fig. 5.1). Other drivers such
as climate warming and nitrogen deposition may exert lesser impacts on

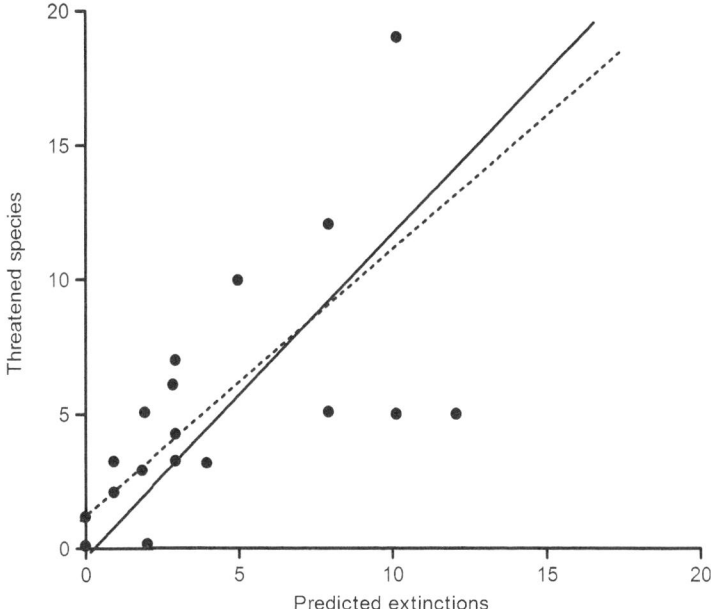

Figure 5.2 The predictions of bird species expected to become extinct following deforestation match the numbers of threatened bird species in insular Southeast Asia. The regression between the points (dashed line) and the line where the two sets of numbers are equal (solid line) is not significantly different. (Reproduced with permission from Brooks *et al.* 1997.)

tropical biodiversity (Fig. 5.1). However, it is highly probable that land-use changes act synergistically with other drivers for accumulative impact on the tropical biodiversity.

However, how do the predicted scenarios of biotic extinctions fit for Southeast Asia? Brooks *et al.* (1997) using the species area equation showed that the number of bird extinctions predicted to occur in Southeast Asia due to deforestation matched closely with the number of threatened species (Fig. 5.2). This relationship shows that a large proportion of threatened bird species will most likely become extinct from Southeast Asia should the current rate of deforestation continue. This prediction may sadly become a reality as the deforestation rate in Southeast Asia is accelerating (see Chapter 1).

Brook *et al.* (2003) reported biotic extinctions from Singapore and used these to project the future biotic losses from Southeast Asia (see also Chapter 3). Singapore has experienced exponential population growth, from approximately 150 subsistence-economy villagers in the early 1800s to four million people in 2002 (Corlett 1992; WRI 2003). In particular, Singapore has transformed itself from a developing country of squatters and slums to a developed metropolis of economic prosperity within the past few decades and, thus, has been widely regarded by the leaders of regional developing countries as the ideal economic model.

However, the success of Singapore came with a hefty price, paid for by its biodiversity. Since the British first established a presence in Singapore in 1819, more than 95% of the estimated 540 km^2 of original vegetation cover has been entirely cleared (Turner *et al.* 1994; Corlett 2000), initially from the cultivation of short-term cash crops (e.g. gambier, *Uncaria gambir* and rubber, *Hevea brasiliensis*) and subsequently from urbanisation and industrialisation (Corlett 1992). Brook *et al.* (2003) showed substantial rates of documented (observed) and inferred (based on what could have occurred in Singapore before habitat loss) extinctions, with most extinct taxa (34–87%) being species of butterflies, fish, birds and mammals (Fig. 5.3). Similar environmental scenarios are already unfolding on a much larger scale in other Southeast Asian countries, such as Indonesia (Jepson *et al.* 2001).

What will be lost, and how fast?

Brook *et al.* (2003) provided some simple extrapolations of the future fate of biodiversity in Southeast Asia, based on the species–area model calibrated to the extinctions observed and inferred to have occurred in Singapore over the past 180 years. They concluded that the current rate of habitat destruction in Southeast Asia will result in the loss of 13–42% of regional populations (upper and lower bounds based on observed/inferred extinction data) of all species by the turn of the next century (Fig. 5.3), of which at least 50% of these could represent global species extinctions, given that approximately half of the biota is endemic to the region (see Chapter 2 and analyses below).

Here we expand upon these broad projections using our unpublished manuscript to provide more specific and accurate information on: (a) how these projected losses will unfold over time; (b) for the well studied taxonomic groups, how many species, in terms of absolute numbers, are we likely to lose; (c) a more detailed taxonomic breakdown of predicted global species losses that result from the loss of regional populations; (d) a decomposition of deforestation rates and predicted losses on the basis of political units (countries) and biological centres of special conservation concern ('hotspots'); and (e) a partitioning of habitat losses into lowland and montane areas, the latter being more prone to other global drivers such as anthropogenically induced climatic warming. The results, in terms of the proportion of local species populations and global species extinctions,

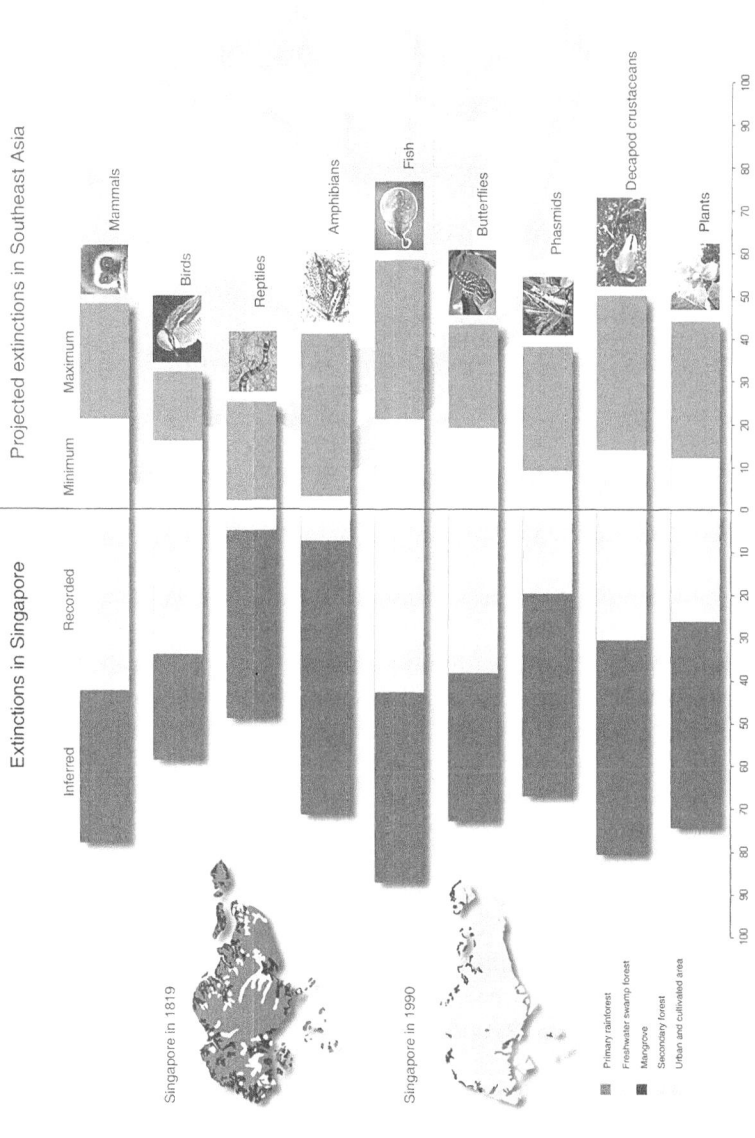

Figure 5.3 Population extinctions in Singapore and Southeast Asia. Black and pale grey bars represent recorded and inferred extinctions in Singapore, respectively. White and dark grey bars represent minimum and maximum projected extinctions in Southeast Asia, respectively. (Reprinted from Sodhi *et al.* 2004b. With permission from Elsevier.) For more details see Brook *et al.* 2003.

Table 5.1 *Estimated original forest cover, remaining forest cover in the year 2000, and the percentage representation of forest of different elevation types (lowland = <1200 m, upland = 1200–1800 m, montane = >1800 m), in each country of Southeast Asia*

| Country | Forest area (000 ha) | | Percentage by elevation | | |
	Original cover	2000 cover	Lowland	Upland	Montane
Indonesia	181 157	91 134	78.6	10.0	11.4
Myanmar	65 755	33 519	87.1	12.9	0.0
Thailand	51 089	13 452	79.5	19.8	0.8
Malaysia	32 691	17 107	86.2	12.8	1.0
Vietnam	32 452	4495	65.9	34.1	0.0
Philippines	28 416	5015	82.1	16.5	1.4
Laos	23 057	11 562	92.1	7.2	0.7
Cambodia	17 652	2405	97.0	3.0	0.0
Brunei	527	267	97.2	2.8	0.0
Singapore	54	0.2	100.0	0.0	0.0
Total	432 850	178 956	81.9	12.1	6.0

actually turn out to be more pessimistic than those estimated by Brook *et al.* (2003), because of the more detailed information used to characterise localised loss of habitat through deforestation. Data sources for the projections presented in this chapter were derived from the World Resources Institute (http://www.wri.org/), Food and Agriculture Organization, Forestry (http://www.fao.org/forestry, FRA 2000 Reports), Conservation International Biodiversity Hotspots (http://www.biodiversityhotspots.org/), World Wildlife Fund (http://www.worldwildlife.org/wildworld/) and other sources (Iremonger *et al.* 1997; Ong 2000; Brook *et al.* 2003; Sodhi *et al.* 2004b).

The past and current state of Southeast Asia's forest areas are summarised in Table 5.1 on a country-by-country basis, and includes both pristine and secondary rain forest, dry forest, mangroves, freshwater swamps, evergreen, semi-deciduous and deciduous broadleaf woodlands, and needle-leaf woodlands, but excludes degraded and exotic forests. This table also provides an indication of the relative representation of lowland (<1200 m above sea-level), upland (1200–1800 m) and montane forest areas (>1800 m) based on the contemporary (year 2000) distribution of residual forest. Because lowland forests have historically suffered substantially greater deforestation than upland and montane areas, due to easier accessibility for swidden farming and timber extraction, and generally more fertile soils (see Chapter 3), these relative proportions are probably biased in respect to areas of high elevation if one considers the original forest distribution.

Table 5.2 *Estimated percentage rate of deforestation and projected future areas of remaining forest for each country of Southeast Asia, and alternatively, for the four major biodiversity 'hotspots' located in this region*

Locality	Deforestation rate/yr (%)	Projected Forest Area (000 ha)			
		2025	2050	2075	2100
Indonesia	− 1.50	62 458	42 805	29 336	20 105
Myanmar	− 1.50	22 972	15 744	10 790	7 395
Thailand	− 2.90	8197	3928	1882	902
Malaysia	− 1.40	9456	6647	4673	3285
Vietnam	− 0.30	4652	4315	4003	3714
Philippines	− 2.10	1415	832	490	288
Laos	− 0.50	3966	3499	3086	2723
Cambodia	− 0.60	9947	8558	7362	6334
Brunei	− 0.30	248	230	213	198
Singapore	0.00	0.2	0.2	0.2	0.2
Sundaland	− 1.49	53 545	36 808	25 303	17 394
Indo-Burma	− 1.53	47 242	32 092	21 800	14 809
Philippines	− 2.10	1415	832	490	288
Wallacea	− 1.50	18 737	12 841	8801	6031
Total	− 1.40	123 310	86 557	61 835	44 942
% Remaining	—	28.5	20.0	14.3	10.4

Current estimated rates of deforestation by country, and according to Southeast Asia's four major biodiversity 'hotspots', are presented in Table 5.2. These rates are based on reported forest losses over the period of 1990–2000, and may very well underestimate the true extent of habitat conversion (WRI 2000). Nevertheless, these figures provide a tractable basis from which to estimate the likely future of Southeast Asia's natural habitats should deforestation continue unabated at its current rate. The results suggest only around 20% of the region's forests will remain intact by 2050, and just over 10% by 2100. Perhaps even more worryingly, only 28.5% will be left by 2025, just two decades from the present. The slowing amount of absolute area deforested is a manifestation of an assumed proportional utilisation of a diminishing forest resource, rather than a fixed area of habitat being deforested each year. On an individual country basis, nations such as Cambodia and Brunei may very well retain a reasonable proportion of their forest area (36% and 38%, respectively, by 2100), whereas for those nations currently suffering severe rates of habitat loss, such as Thailand and the Philippines, the projected future is bleak, with almost no forest predicted to be left by the next century (1.8% and 1.0%, respectively).

Turning to the issue of how this large-scale destruction of habitat is likely to impact upon the region's biodiversity, Tables 5.3 and 5.4 summarise the projected local and global losses of populations/species across the region by the year 2100, broken down by country, biodiversity hotspot and taxonomic group. These projections rely on the application of the species-area model derived by Brook *et al.* (2003), which was based on a detailed historical analysis of extinctions following deforestation of Singapore, calibrated to lower (observed extinction) and upper (inferred extinction) bounds of uncertainty for each taxonomic group. Combining this modelling framework with information on the current known species composition, richness, and levels of endemicity for each country and hotspot in Southeast Asia (see also Sodhi *et al.* 2004b) and the original, current and projected areas of forest habitat for each respective locale (detailed in Tables 5.1 and 5.2), it is possible to (a) reconstruct the full (albeit approximated) species composition for each area prior to any anthropogenic intervention, and (b) take the estimates of regional percentage loss of local populations presented in Brook *et al.* (2003) one step further, by estimating the actual number of species likely to be either extinct or consigned to extinction, both locally and globally (applies to regional endemics), in the future. By the year 2100, only 10% of the region's original forests are likely to remain intact (Table 5.3 and 5.4).

Above and beyond the direct impact of deforestation on the biota, other aspects of anthropogenic change, such as the shifting altitudinal range of upland and montane species associated with global warming (Grabherr *et al.* 1994), will be likely to further deplete the biodiversity of the region, especially given the relatively high proportion of elevated areas (see Table 5.1) on the large tectonic islands such as Java and Sulawesi that are characteristic of the Indonesian archipelago, where over one-third of the forested area is found at altitudes exceeding 1000 m and maintain very high levels of species endemicity due to their isolated nature and varied biogeographical history (see Chapter 2). For instance, the regions of Peninsular Malaysia, Borneo, Sumatra, Java and Sulawesi have a combined total of 212 birds, mammals, reptiles and anurans, or 28% of their total species count for these groups, found only in these montane forests (Ong 2000). A substantial contraction of the climate zones associated with these high elevation areas due to a warming climate will clearly place additional pressure on these species, and will in all probability, result in further extinctions not associated directly with deforestation.

In summary, these results suggest that anywhere from 23%–79% of local populations will be extinct or consigned to extinction within the next century, a figure exceeding the coarser estimate of 13%–42% by Brook *et al.* (2003). Even more worryingly, an almost equal proportion of endemic

Table 5.3 *Projected loss of local populations by the year 2100 for each country in Southeast Asia, based on the known (minimum) and inferred (maximum) number of resident species' populations, the projected area of remaining forest by the turn of the next century, the current rate of habitat destruction, and the upper and lower bounds of the scaling (z) parameter of the species-area curve derived for each taxonomic group in Southeast Asia by Brook et al. (2003).*

Country	Number of local species losses by taxonomic group and country (minimum and maximum)															
	Mammals		Birds		Reptiles		Amphibians		Plants		Butterflies		Freshwater fish		Total	
Indonesia	161	537	231	624	32	407	12	251	5797	28 057	—	—	1333	5399	7566	35 275
Myanmar	93	309	77	206	11	142	3	71	1374	6617	—	—	91	368	1650	7714
Thailand	132	410	116	295	26	291	8	139	3861	16 560	—	—	159	605	4303	18 299
Malaysia	97	360	66	192	17	232	9	203	3182	16 936	—	—	125	573	3495	18 497
Vietnam	66	447	64	299	12	249	6	221	2046	19 303	—	—	83	802	2277	21 322
Philippines	83	484	182	695	24	369	10	280	3288	24 586	—	—	535	4691	4122	31 105
Laos	52	315	51	219	5	115	2	87	1594	13 430	—	—	16	130	1722	14 295
Cambodia	20	75	23	69	2	34	0	46	—	—	—	—	22	103	67	285
Brunei	24	120	11	42	1	25	0	3	560	4102	—	—	10	66	607	4356
Singapore	19	72	74	129	6	105	2	60	594	4272	145	482	26	208	866	5328
Total	746	3129	895	2770	133	1969	52	1320	22 298	129 591	145	482	2400	12 944	26 529	151 723
% local loss	32.7	82.2	29.1	67.3	5.1	58.6	5.3	81.8	22.4	78.9	38.1	55.9	36.3	91.7	23.0	79.4

141

Table 5.4 *Projected loss of global species by the year 2100 for each country and the four major biodiversity 'hotspots' in Southeast Asia, based on the known (minimum) and inferred (maximum) number of resident species endemic to that geographical unit, the projected area of remaining forest by the turn of the next century, the current rate of habitat destruction, and the upper and lower bounds of the scaling (z) parameter of the species-area curve derived for each taxonomic group in Southeast Asia by Brook et al. (2003)*

Country	Mammals		Birds		Reptiles		Amphibians		Plants		Butterflies		Freshwater fish		Total	
Indonesia	97	323	149	402	13	167	10	219	3454	16 715	96	319	149	603	3967	11 988
Myanmar	1	3	1	3	2	20	1	24	271	1304	1	4	49	199	326	991
Thailand	2	6	1	3	3	32	2	28	625	2681	19	57	45	173	697	1709
Malaysia	2	7	5	14	3	43	3	73	1393	7413	34	125	19	86	1458	4457
Vietnam	1	4	2	11	2	41	2	96	246	2316	57	365	26	256	336	1004
Philippines	61	354	84	320	15	228	7	197	2345	17 533	202	1086	61	537	2774	6342
Laos	0	2	0	1	0	1	0	3	9	73	2	11	34	280	45	113
Cambodia	0	0	0	0	0	0	0	3	1	4	0	0	0	1	1	5
Brunei	0	0	0	0	0	0	0	0	1	5	4	19	0	4	4	12
Singapore	1	1	1	1	0	0	0	0	0	0	0	0	0	0	2	2
Sundaland	36	124	35	96	12	150	8	166	2995	14 756	–	–	–	–	3085	15 291
Indo-Burma	25	95	39	115	10	134	6	127	1562	8294	–	–	–	–	1642	8765
Philippines	60	351	88	335	15	231	7	200	2147	16 055	–	–	–	–	2317	17 171
Wallacea	38	128	62	167	5	67	2	32	296	1433	–	–	–	–	403	1826
Total	164	701	243	755	42	582	26	643	8343	48 043	414	1986	384	2135	9611	43 054
% endemic loss	37.5	85.0	29.6	65.7	6.3	65.2	5.1	77.7	23.7	80.5	36.1	85.0	36.6	90.8	24.1	63.3

Number of global species losses by taxonomic group and country (minimum and maximum) species (24%–63%) will be likely to perish, and therefore be counted amongst the irreversible tally of global species extinctions (see IUCN 2004 for a full listing of current known extinctions). In raw numbers, these proportions of global species extinctions translate to between 859–4815 vertebrate species and 8343–48 043 species of vascular plants (Table 5.4). Moreover, this longer-term projection tends to mask the fact that the vast majority of these 'losses', especially in terms of species consigned to extinction (e.g. below their minimum viable population size, or alternatively, functionally extinct with respect to their original role in the ecosystem), may actually unfold much more rapidly due to the non-linear absolute rate at which habitat is being lost (much of the forests will have been cleared by 2025). Functional extinction of vertebrates in the region by 2025 may be as high as 3954 species, and 39 113 vascular plants, associated with the loss of 72% of the original (pristine) habitat (Table 5.2). Thus even when one acknowledges the uncertainties surrounding these figures, and the admittedly large margins between the lower and upper estimates of predicted species losses, these figures remain of grave concern to the future of the region's and the globe's biodiversity, and ultimately, the welfare of humans who depend upon healthy forests for their livelihoods and survival (see Chapter 6).

Summary

1. Humans are directly or indirectly accelerating the rate of biotic extinctions.
2. We show that with the current rate of deforestation, only 10% of the existing Southeast Asian forests are likely to remain by the year 2100.
3. Such a predicament is likely to result in the loss of 23%–79% of local (national) populations by the year 2100, or up to 150 000 total populations.
4. Similarly, 24%–63% of endemic species may also perish during the same period, representing the global extinction of up to 4000 species of vertebrates and almost 40 000 species of vascular plants.

Chapter 6

Challenges and options for conservation in the region

The major challenges to overcoming the present and imminent threats to Southeast Asian biodiversity are primarily socioeconomic in origin, including poverty, chronic shortage of conservation resources (e.g. expertise and funding) and weak national institutions. As the regional societies strive expeditiously to match the living standards of developed nations, environmental issues are inevitably marginalised (Bawa & Dayanandan 1997). Yet despite this pessimistic outlook, there are still conceivable ways to conserve at least some of the regional natural resources for future generations. Given that many of the drivers of biodiversity loss (e.g. international demand for rain forest timber, climate change) are issues that transcend national boundaries, any realistic solution will need to involve a multi-national and multi-disciplinary strategy, including political, socioeconomic and scientific input, in which all major stakeholders (governmental, non-governmental, national and international organizations) must partake (Balmford et al. 2002). Below we discuss some of the major challenges faced by conservation biologists in Southeast Asia.

Create public awareness, empower local people

At the local level, scientists, non-governmental organizations and governmental agencies should recognise the possibility and importance of attempting to sustain economic development without needing to obliterate their natural resources. There seems to be a general ignorance or apathy about environmental issues in the Southeast Asian public arena (Whitten et al. 2001). For instance, only 27% of 74 people surveyed in Indonesia thought that conservation of species was of major concern (Jepson 2001). More worryingly, this percentage declined to 15% when asked about the necessity of preserving native forests and other wild places. It is imperative to convince the populous and governments of the region that biodiversity matters (Kinnaird & O'Brien 2001; see Chapter 1).

We suggest that through activities such as intensive public education programmes, awareness should be raised about the severity and possible short- and long-term ramifications of the looming disaster facing Southeast Asia's biodiversity. The media could be engaged creatively for this purpose. For example, local radio and television could be used as communication vehicles for spreading this message to both the older generations, through local languages and dialects, and future generations, through cartoons. This media campaign could also be extended internationally to canvass for much needed conservation funds from foreign sources. Lack of a truly free press in some Southeast Asian countries may hinder public awareness campaigns but this problem can be countered using global communication mediums such as the internet.

As mentioned, it would be wise to include environmental awareness in the education of children. They should be given opportunities to visit intact natural areas and observe wild animals (Liow 2000). Without such measures, there is a real danger of developing a general alienation towards environmental issues in youths and adults, as has been noted in heavily urbanised Singapore (Kong et al. 1997; Kong et al. 1999). Worse still, they could become environmental destroyers (i.e. 'lost generation') rather than protectors early in life (Fig. 6.1).

One excellent example of a positive public education campaign occurred in eastern Malaysia (Bennett et al. 2000). The Sarawak Forest Department has apparently been attempting to educate local people about environmental issues for a number of years. Innovative methods such as dance and role playing are being used to educate people about wildlife conservation in a stimulating way. In addition, people are being instructed in how to run small businesses, attract tourism dollars, and are taught the English language in the hope that they can obtain jobs in the tourism industry and other related enterprises. We feel that this is a superb, albeit rare example of what can be achieved, and should be mimicked by relevant national and international organizations in other Southeast Asian locales.

We concur with Whitten et al. (2001) that local non-governmental organizations (NGOs) remain a valuable but often underutilised hope for the biodiversity conservation in various Southeast Asian countries. The local people working in these organizations can also play important roles in compiling biodiversity inventories and assisting in the development of sound environmental impact assessments (Janzen 2004; Sheil & Lawrence 2004). However, organizations should be cautious that certain NGOs might be infiltrated by the people with their own economic and political agendas (Bryant 2002). Additionally, NGO staff often lack biological training (e.g. in proper scientific sampling methods, data analyses and interpretation) and

Figure 6.1 The endemic yellow-sided flowerpecker (*Dicaeum aureolimbatum*) killed (top) using sling shot by 6–8 year old children (bottom) in Sulawesi (Indonesia).

efforts should be made to rectify this deficient expertise and avoid duplication of effort (Mace *et al.* 2000). Further, training of students in conservation biology and of forestry guards against improper management practices, will certainly help to facilitate good conservation outcomes (Kinnaird & O'Brien 2001; Bawa *et al.* 2004a).

It has also been suggested that national and local NGOs should work in close collaboration with international NGOs, to ensure that the perpetrators of environmental destruction are prosecuted (Robertson & van Schaik 2001). It is all too often the case that offenders walk free due to lack of funds to pursue legal proceedings.

Foster international collaboration

All developing tropical countries, both within and beyond Southeast Asia, should be assisted by the developed world in establishing better environmental practices (see Aldous 2004). This can be facilitated by fostering stronger collaborations among national, regional and international stakeholders. Some such partnerships already exist between organizations such as ADB (Asian Development Bank), ASEAN (Association of Southeast Asian Nations) and UNEP (United National Environmental Program). Existing partners should be strengthened, and newly forged relationships developed, so as to maximise international assistance for poorer countries in improving their environmental performance. Environmental indicators (e.g. the state of protected areas) for each country should be collected rigorously across all countries in the region (http://www.adb.org). If possible, nations can be subjected to periodic environmental audits, conducted under the auspices of an impartial international body (e.g. the United Nations) and mediated by an international panel of experts.

Local and national NGOs can also be used to evaluate degradation of habitats and the adequacy of biodiversity protection. Environmental audits should aim to assess objectively and quantitatively the level and efficacy of protection of each country's natural habitat and biodiversity. The results of such audits should preferably be tied to appreciable consequences. Countries with poor conservation practices should not be assisted monetarily with their economic development (excluding humanitarian aid); while countries that have become more environmentally conscious and friendly through time could be rewarded. International development loans and grants would ideally be tied to the soundness and sustainability of environmental practices (Kremen *et al.* 2000; Robertson & van Schaik 2001). Only with such measures in place will there be the political will to better protect biodiversity. We admit that this approach may initially result in some environmental destruction and international ill-feeling, but can see few other avenues for realising tangible biodiversity protection in regions such as Southeast Asia over the long-term. Yet alternatives such as a blanket ban on the national export of timber from nations which allow unsustainable forest extraction would be likely to be ineffective, difficult to enforce, create black markets, and stymie initiatives that have the potential to foster conservation management (Jepson *et al.* 2001).

Eradicate corruption

High levels of political corruption exist in some of the developing countries in tropical regions (Talbott & Brown 1998; Laurance 2004). Political corruption can hinder conservation by reducing effective funding for conservation and distorting priorities through the overexploitation of forests, wildlife, fisheries and other natural resources. Recently, pervasive corruption has been shown to promote rampant illegal logging in Kalimantan (Indonesian Borneo) (Smith *et al.* 2003) and the Philippines (Kummer & Turner 1994). In Borneo, timber entrepreneurs apparently bribed local authorities to issue excessive numbers of short-term logging contracts. Local tribal leaders, military, police and provincial forestry officials were also bribed (Smith *et al.* 2003). To facilitate the export of wood to Malaysian Borneo, border patrols and Malaysian officials also were paid off. This has resulted in a catastrophic loss of much of the island's forest. This example dictates that efforts to reduce or eradicate the effects of corruption are urgently required if effective conservation is to be achieved.

A large part of the illegal logging money is laundered through the banks. Realising this problem, in 2003 Indonesia took the step of imposing extremely heavy penalties for laundering the proceeds of forestry and environmental crimes, with a maximum penalty of a 15-year jail-term and a fine of US$1.7 billion (CIFOR 2003). Stockmarkets also need to be vigilant so that money is not invested into companies guilty of unsustainable timber practices, enforced via the requisite use of certificates of sustainability for timber products (Barr 2001). Laurance (2004) suggested that developing nations should be assisted by the governments of developed countries in their efforts to curb corruption.

Enhance protection

Globally, 10–27% of threatened vertebrates do not fall under the umbrella of any legally protected areas (Rodrigues *et al.* 2004). Protected areas (reserves or national and regional parks) may be the only hope for retaining

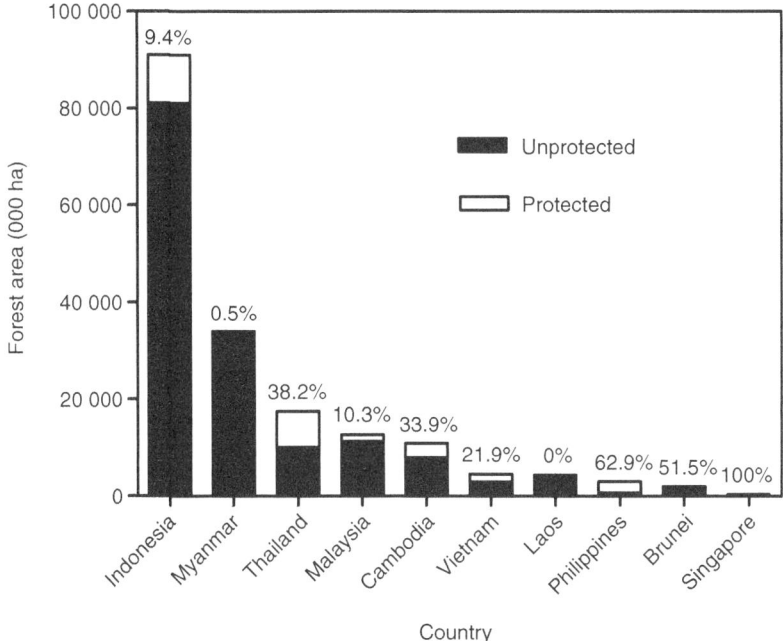

Figure 6.2 Forest protection in Southeast Asia. Numbers on bars represent percentage of forest cover that is legally protected. (From http://www.unep-wcmc.org/forest/data/cdrom2.)

a reasonable proportion of Southeast Asian and tropical biodiversity (Bruner *et al.* 2001). Yet at present, only 16% of the remaining forests in Southeast Asia are protected (Fig. 6.2). It is particularly worrying that much of the remaining forests (84%) in Southeast Asia are not only unprotected, but also situated in countries that do not have the resources to protect them adequately (Fig. 6.2; see Fig. 1.7). Currently, less than 5% of the costs needed for effectively maintaining protected areas are being met in developing Asia, and of an estimated total of US$6 billion spent each year on managing protected areas, <12% is spent in developing areas (Balmford *et al.* 2003b; Fig. 6.3). The current trend in Southeast Asia suggests that forest loss will be likely to increase with economic growth and human population density (see Chapter 1, Fig. 1.7). Even fewer of the critical headwaters of the catchments of many of such forested areas are protected.

Forested areas with high conservation value (e.g. 'hotspots' rich in endemic species; Myers 2000) could be purchased and protected in a competent manner by international agencies through collaborations with national

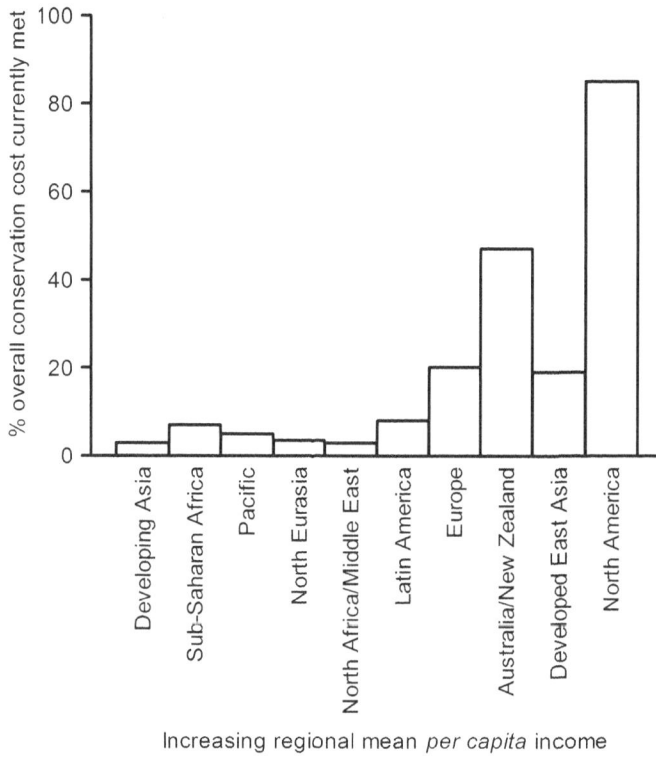

Figure 6.3 Regional variation in the percentage of the overall cost of effective reserve networks that are met. These figures refer to the estimated overall costs of expanded networks (from James *et al.* 2001), but the positive correlation with mean regional gross national product (GNP) holds also for the percentage of existing reserve management costs that is currently met ($r_s = 0.72$, $n = 10$, $P < 0.05$). (Reprinted with permission from Balmford *et al.* 2003b.)

organizations (James *et al.* 1999; Whitten *et al.* 2001). This strategy would not only provide immediate protection for the native biodiversity but also establish a mechanism to sustain the effort. A reasonable proportion of the funding required for such endeavours could come from developed countries, both in the region and beyond (Balmford & Whitten 2003). Involvement of private-sector institutions and private philanthropists in biodiversity conservation (e.g. purchase of land for biodiversity reserves) needs to be explored more aggressively (Daily & Walker 2000; Balmford & Whitten 2003). Further, protected areas should also encompass poorly

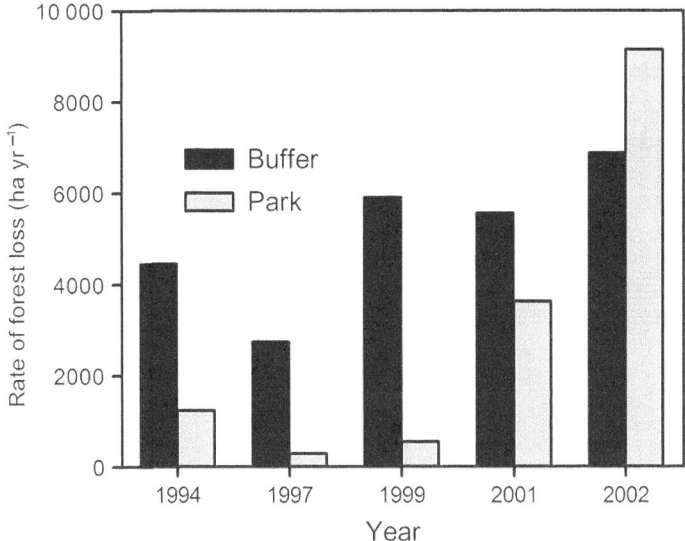

Figure 6.4 Annual rates of forest cover change in the buffer zone and within the Gunung Palung National Park (Kalimantan) from 1988 to 2002, from a 1988 baseline. During this period, a total of 628 km^2 of lowland dipterocarp forest within the park buffer and 257 km^2 within the park interior were lost, primarily through logging and conversion to plantations. After 1999, when approximately 75% of the buffer had been deforested, the rate of deforestation within the park increased to 9.5% year^{-1}. (Reprinted with permission from Curran *et al.* 2004. Copyright 2004 American Association for the Advancement of Science.)

studied and unexplored natural habitats, as well as different habitat types (e.g. mangroves, swamps) and elevations (montane areas).

About 23% of existing tropical humid forests are officially protected worldwide (Chape *et al.* 2003). However, it is not known if these reserves receive adequate protection against activities such as illegal logging and poaching. Some so-called 'protected' forests in Southeast Asia have become isolated, degraded and/or deforested (Whitten *et al.* 2001; Curran *et al.* 2004). Protected lowland forests of the hyper-biodiverse region of Kalimantan have declined by 56% between 1985 and 2001, due primarily to intensive logging (Curran *et al.* 2004). This forest decline is not restricted to the parks, also occurring within the buffer areas (Fig. 6.4). As regenerating forest shows some potential for biotic recovery in certain areas (see Chapter 3), buffers could serve as excellent reservoirs to extend park boundaries.

Other studies illustrate similar problems faced by protected areas throughout Southeast Asia. Many protected areas in this region suffer

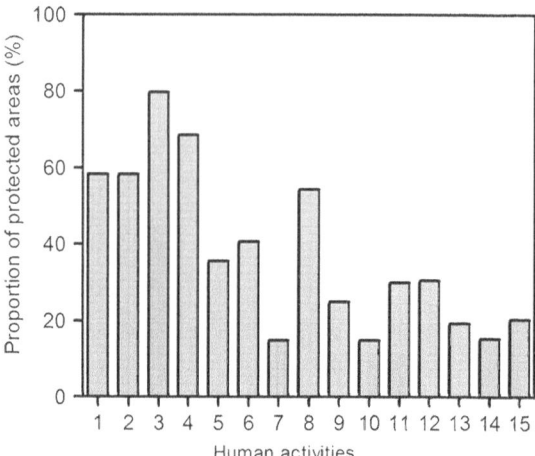

Figure 6.5 Proportion of protected areas in Myanmar with human activities. 1 = hunting for subsistence and wildlife trade; 2 = fuel-wood collection; 3 = extractions of non-timber forest products (orchids, palm leaves, grass, rattan, honey, mushrooms, bamboo, resin from dipterocarps and medicinal plants); 4 = grazing by domestic cattle, sheep and horses; 5 = fishing (crabs, prawns and fish); 6 = shifting cultivation; 7 = mining; 8 = permanent human settlements; 9 = roads and railway lines; 10 = plantations of sugar cane, rubber and oil palm; 11 = military camps and/or insurgents, indicating availability of firearms; 12 = permanent cultivation; 13 = tourism; 14 = breeding centres for ducks, fish and other animals; and 15 = extractions of timber species such as teak. (Data from Rao *et al.* 2002.)

from three main threats: illegal logging, encroachment by shifting cultivators, and fires (http://www.fao.org). In Indonesia (Pulau Kaget Nature Reserve), excessive infringement by expanding farms has resulted in the loss of habitat for the threatened proboscis monkey (*Nasalis larvatus*). Translocation of these monkeys resulted in their ultimate demise from this reserve, and did not help them establish new populations elsewhere because it was ill-planned (to unprotected areas) and poorly executed (13 monkeys died during capture) (Meijaard & Nijman 2000).

Likewise, a large proportion of the protected areas of Myanmar suffer from excessive activities such as extraction of non-timber products (Rao *et al.* 2002) (Fig. 6.5). Insufficient numbers of poorly trained staff are left to manage the parks of Myanmar (e.g. prevent poaching), but cannot do so in an adequate manner (Rao *et al.* 2002). For tropical protected areas, the density of guards has been shown to correlate positively with the effectiveness of the park protection (Bruner *et al.* 2001).

Table 6.1. *Potential future biodiversity losses in Singapore with the loss of protected areas*

Taxon	% (n) of species restricted to Singapore's protected areas*	Total % extinct (increase) with loss of protected area[†]
Decapods[§]	81 (13)	87 ($\times$ 2.9)
Phasmids	100 (33)	100 ($\times$ 5.1)
Butterflies	63 (149)	77 ($\times$ 2.0)
Fish[§]	60 (21)	77 ($\times$ 1.8)
Amphibians	76 (19)	78 ($\times$ 10.5)
Reptiles	50 (59)	52 ($\times$ 10.7)
Birds	8 (12)	39 ($\times$ 1.1)
Mammals	46 (12)	69 ($\times$ 1.6)
Weighted Mean	50 (312)	66 ($\times$ 2.1)

From Brook *et al.* (2003).
*Reserves are Bukit Timah (71 ha), Nee Soon (935 ha), MacRitchie (484 ha), Lower Pierce forest (50 ha) and Botanic Gardens (7 ha). Numbers in brackets are total number of species per taxon restricted to the reserves.
[†]Calculated as (# species extinct = # species restricted to reserves)/original # of species. 'Increase' = (# species extinct = # species restricted to reserves)/# species extinct; e.g. '(x 2.0)' signifies a two-fold increase in the extinction rate. Based on recorded extinctions only, so represents a minimum estimate of biodiversity loss.
[§]Freshwater only (excludes marine species).

We are sure that Myanmar is not an anomaly in terms of the state of Southeast Asian protected areas. Yet even so, protected areas nevertheless remain a major hope for biodiversity conservation. In their analysis of 93 protected areas in 22 tropical countries, Bruner *et al.* (2001) found that protected area status may be succeeding in preventing land clearing; 43% of them had no net land clearing since their establishment. They recommended that management actions such as genuine law enforcement, boundary demarcation, and direct compensation to local communities, can assist in better protection of parks and thus of tropical biodiversity. Establishment of appropriate protected areas is also critical in halting future extinctions. Brook *et al.* (2003) show that the already alarming rate of biotic extinctions in Singapore will increase by 66% with the loss of its few remaining protected areas (Table 6.1).

Therefore, the security of tropical biodiversity can be augmented effectively by creating new protected areas in biologically important sites, diverting more funds to improve the management of existing protected areas, and perhaps by creating links (habitat corridors) among the protected areas (Bowles *et al.* 1998); see below for further details. It is important that new protected areas are identified correctly for optimal conservation outcomes. For example, most, if not all areas of high bird endemism should be protected (Norris & Harper 2004). Threatened endemic bird species are likely to be habitat specialists and as such are particularly vulnerable to habitat disturbance: clearly a high priority for conservation (Norris & Harper 2004). In the tropics, the extent of current protected areas does not correspond particularly well with the number of threatened species (Kerr & Burkey 2002). Effective forest reserves should also be of sufficient area to maintain viable populations of its resident threatened species, as mere presence does not guarantee long-term survival. Monitoring of ecological data such as population sizes and demographic rates will be needed to evaluate adequately the population viability of at least the keystone species of a protected area (Caughley & Gunn 1986).

Because of the typically patchy distributions among many tropical taxonomic groups (e.g. birds), care should be taken to consider the complimentarity and comprehensiveness of the species composition of proposed future reserves as part of the larger protected areas network (Diamond 1980; van Balen 1999; van Balen *et al.* 1999). A crucial part of this process is the consultation of lists of threatened species and their distributions, because a large reserve may fail to protect all or even the most vulnerable species. For example, 83% of 29 endangered butterfly taxa do not occur in any of 14 existing or proposed protected terrestrial areas in the Philippines (Danielsen & Treadaway 2004). In addition to reserve location, reserve size is also critical. Small reserves may: 1) not contain sufficient volume or diversity of resources, 2) permit the infiltration and spread of parasites and invasive species from the surrounding matrix, and 3) only contain tiny populations of many species, leaving them vulnerable to the perils of environmental stochasticity (e.g. localised wildfires, hurricanes). In fact, even a 198 ha remnant failed to encompass the ranges of vulnerable species such as the black-necked aracari (*Pteroglossus aracari*) and spot-billed toucanet (*Selenidera maculirostris*) near Lagoa Santa, Brazil (Christiansen & Pitter 1997). Although large protected areas are clearly often required, there is no consensus about the minimum reserve sizes in the tropics. Leck (1979) recommended tropical reserves be at least as large as 20–30 km^2, but preferred those >100 km^2. Only large fragments ($\geq$100 km^2) are probably capable of supporting intact populations of many vertebrate fauna with large home range requirements (Curran *et al.* 2004). Other authors have also suggested that reserves of several thousand square km may be needed in the tropics to halt or diminish the chances of mass extinctions (Terborgh 1974; Whitmore 1980; Myers 1986; Thiollay 1989; Terborgh 1992). Such large reserves may be possible in areas of Southeast Asia such as Myanmar, where extensive

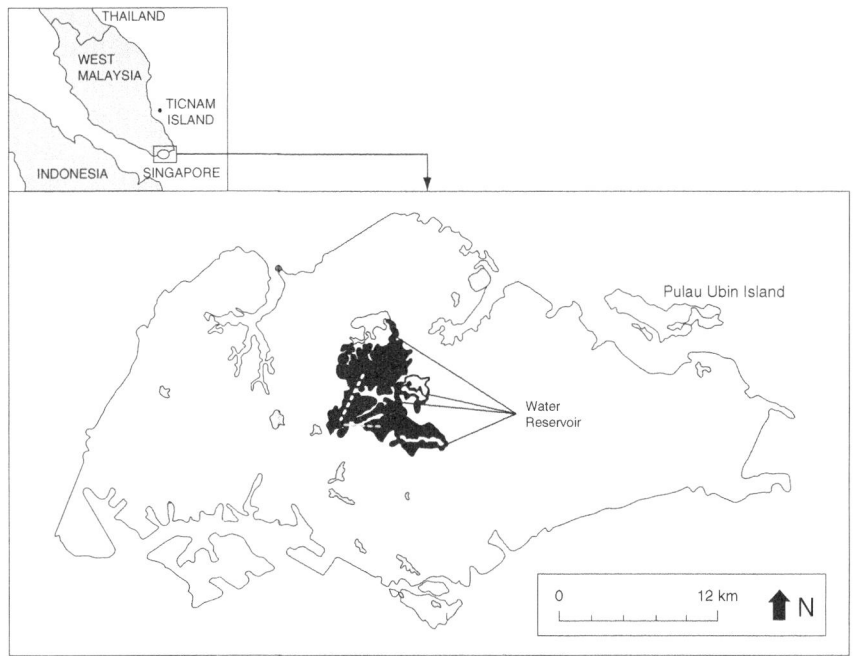

Figure 6.6 A regional map of Southeast Asia with Singapore's location. Black patches represent forests in the Central Catchment Nature Reserves of Singapore. Grey patches represent forest fragments. Dashed lines represent proposed connections between forest patches for the avifauna. (Reprinted from Castelletta *et al.* 2005). With permission from Elsevier.

undisturbed forests still exist (Rao *et al.* 2002). In areas where large forested tracts are unavailable, forests around the existing reserves could be restored (Whitmore & Sayer 1992).

The abundance of potential predators should not be unnaturally high in reserves (Terborgh *et al.* 1997). In addition, mature and/or high quality forests should ideally be protected. Sodhi (2002) found that reserves composed of poor quality mature forest (i.e. heath forest) contained fewer rare bird species than those with high quality but otherwise similar-sized mature forests. This result suggests that some rare species may need high quality forest with a greater diversity of ecological niches.

Habitat links (e.g. fence rows and windbreaks) between forest patches or reserves may facilitate dispersal and buffer against local extinctions and population declines in fragmented landscapes (Lovejoy *et al.* 1997; Sekercioglu *et al.* 2002). Data on efficacy and adequacy of habitat corridors in the tropics are limited, however. High vegetation cover (both native and non-native) can enhance the attractiveness of corridors for forest species such as the short-tailed babbler (*Malacocincla malaccensis*), thereby enhancing the likelihood of movement through these areas and the re-colonisation and use of such edge habitats (Sodhi *et al.* 1999). Castelletta *et al.* (2005) recommended that patches isolated about 10 years ago in Singapore should be re-connected to facilitate bird movements (Fig. 6.6). On the negative side, corridors can be expensive to create and maintain, may not serve well in heavily fragmented landscapes, and may be counter-productive by facilitating the spread of predators and diseases (Simberloff & Cox 1987; Simberloff 1992).

Promote biodiversity-friendly logging practices

Poor logging practices prevail widely in the tropics (Bowles *et al.* 1998; Laurance 2000; Putz *et al.* 2000). One of the reasons for this unfortunate situation is because 'unsustainable' logging can yield higher profits than sustainable activities, with as much as 450% more money to be made through the former (Bowles *et al.* 1998; Putz *et al.* 2000). Sadly, there seems to be neither the institutional capacity nor political will to counter such a large financial disparity (Smith *et al.* 2003). Certainly, international assistance through environmentally friendly policies such as carbon-credit offset funds and other such means may be needed to promote sustainable logging in the tropics (Kremen *et al.* 2000; Putz *et al.* 2000). Unsustainable logging can benefit certain individuals but if one accounts for losses to humanity (e.g. flood protection and carbon stocks), it results in 14% economic loss compared to sustainable logging (Balmford *et al.* 2002).

In addition to clear-felling, selective logging of rain forests is widely practiced in Southeast Asia, but the way in which such operations are carried out can often be improved substantially to strike a better balance with the needs of biodiversity (Table 6.2). In Malaysia, for example, forestry guidelines dictate that all commercial non-dipterocarp species with dbh $\geq$45 cm, and dipterocarp species with dbh $\geq$50 cm, can be harvested, provided sufficient stems remain in the logging concession for a repeat harvest in about 30 years (Appanah 1998). Similarly, in Indonesia, all commercial trees with dbh of >60 cm can be felled over a 35-year cutting (logging) cycle (Sist *et al.* 2003). Understorey vegetation is removed two, four and six years after logging, to control woody climbers and non-commercial samplings. This treatment is intended to facilitate the growth of commercial tree species, but can be detrimental to those (often species-rich) biotas that depend on the understorey. More importantly, destruction of the

Table 6.2. *Actions needed to make forest harvesting biodiversity-friendly*

1. When selective logging is practiced, reduce the impact by minimising skid trails and damage to the residual vegetation.
2. Determine the maximum number of trees that can be harvested sustainably. For dipterocarp forest, harvest of eight trees per ha or less is recommended.
3. Logging cycles should be greater than 80 years.
4. Assess minimum diameter cutting limits based on the structure, density and diameter at reproduction of the target species.
5. Minimise the size (e.g. $< 600 \, \text{m}^2$) and connectivity of gaps.
6. Refrain from understorey clearance.
7. Set aside pristine areas of appropriate sizes (e.g. >1000 ha) with adequate connections to other such areas within logging concessions.
8. Identify the species of greatest conservation concern (e.g. endangered/threatened species) and ensure that their needs are catered for adequately during logging activities.
9. Convert logged rather than pristine areas into plantations for commercially valuable trees.
10. Prevent excessive human invasion into logged areas. Attempts should also be made to deny access to the poachers of such areas.

Modified from Sist *et al.* (2003).

vegetation of the forest floor can disrupt vital ecological processes such as pollination by decreasing habitat suitability for bees, and reduce the efficiency of nutrient recycling, thereby damaging a forest's capacity to regenerate (see Chapter 3).

The trunk diameter limits to harvesting described above are usually designed to accommodate the market demands and processing technologies, rather than being made with biological considerations in mind. Sist *et al.* (2003) recommended that minimum diameter cuts should actually be limited to >60 cm for dipterocarp trees, to allow them the possibility of reproducing successfully prior to harvesting. Dipterocarps, a key and iconic element of the Southeast Asian lowland forest flora, are renowned for their inter-specific mass fruiting events, usually in cycles of 3–5 years (Appanah 1993; Curran *et al.* 1999; see also Chapter 2). This mass fruiting is thought to have evolved as a mechanism to satiate seed predators such as pigs (*Sus* spp.) and thus allow the significant proportion of fruits to survive and germinate (Curran & Leighton 2000). Therefore mass fruiting is vital to forest regeneration. Similar ecological considerations should be given when deciding minimum diameter cutting limits for other commercial tree species (Table 6.2).

A simulation study showed that the standard 40 year logging cycles may be overly frequent for the forests of Sabah (Huth & Ditzer 2001). They recommended logging cycles of >80 years for the greatest benefit to biodiversity conservation and in minimising the risks of erosion and nutrient loss, whilst still being viable commercially. Such logging cycles should be combined with careful logging practices, including minimising the creation of logging access roads and careless felling of untargeted stems during selective logging. Lengthening the logging cycles will certainly result in lower economic returns (Sist *et al.* 2003), meaning government subsidies may be needed to offset such economic losses. Selective logging can damage >50% of a targeted forest stand (Sist *et al.* 2003). Therefore, reduced impact logging practices (e.g. carefully controlled felling and skidding, and reduced damage to soil and residual trees) has been proposed as an alternative. Reduced impact logging (RIL) has been found to reduce tree injury and death by 18% compared to conventional selective logging in Kalimantan (near Tanjung Redeb) (Bertault & Sist 1997). Clearly, RIL can be very effective in achieving more sustainable forestry, and should be more widely practiced (Kremen *et al.* 2000).

To halt or even dampen the deforestation rates in Southeast Asia and other tropical areas, a prudent measure would be to rely more on plantations for timber and timber products (Durst *et al.* 2004). These plantations can be encouraged especially in already deforested areas, and may serve to provide on-going employment opportunities for local communities. There is also a need to curtail illegal logging that can operate deep within protected areas (Whitten *et al.* 2001). It is worrisome that illegal loggers sometimes hold prominent political posts, and that resident military have also been observed supporting such activities (Kinnaird & O'Brien 2001; Whitten *et al.* 2001). There exists a network of illegal loggers operating not only within Indonesia, which has received most publicity, but also working in close collaboration with industries from Singapore and Hong Kong (Robertson & van Schaik 2001). There have been reports of illegal tropical wood smuggling from Sumatra to Singapore and Malaysia (Anonymous 2004a). Due to the absence of witness protection programmes, activists and forestry officials receive death threats, and may even be killed (Robertson & van Schaik 2001; http://www.ecologyasia.com/NewsArchives/May_2001/thejakartapost.com_20010508.H01.htm). Similarly, in Myanmar and Cambodia, illegal logging has been widely used to fund military operations and may go hand in hand with drug trafficking (Talbott & Brown 1998). Timber certification can be used as a means to restrict illegal logging. However, such certification should ideally also consider impacts of logging on the biota (Bennett 2000). For example, logging companies can be encouraged to retain pristine areas within their concessions via timber certification regulations, and enforce trade bans on the commercial exploitation of wildlife and wildlife products in their own forestry concessions (Robinson *et al.* 1999).

Balance livelihood and conservation

The conservation management of protected areas in Southeast Asia will inevitably cause some disruption to the livelihood of indigenous communities. For example, the creation of nature reserves may be detrimental to local inhabitants who are reliant on the forest for sustenance (e.g. bush meat and localised traditional agriculture), and have few available substitutes at acceptable prices (Milner-Gulland & Bennett 2003). Continued conflicts around certain tropical protected areas indicate that social issues need better consideration for sustainable conservation (Bawa *et al.* 2004b). Since nature conservation is a luxury that hungry stomachs can usually ill afford to indulge, the local communities need to be compensated. But instead of direct financial remunerations that some authors have suggested (du Toit *et al.* 2004), we recommend long-term and sustainable solutions that involve the development of compatible employment opportunities (e.g. resource management) or the introduction of alternative food sources for the impacted communities (see Chapter 4). The due consideration of such socioeconomic forces is paramount to the success of any conservation exercise (Adams *et al.* 2004). In Thailand it has been shown clearly that poachers can be converted into biodiversity custodians with the proper economic incentives (Poonswad *et al.* 2005).

Although controversial, the idea of community-based conservation needs to be explored seriously (Berkes 2004). Community-based ecotourism can economically assist the local communities but such projects may require long-term funding commitment and do not always succeed (Kiss 2004). In some cases, it may be possible for local communities to extract non-timber products sustainably from protected areas (Mendelsohn & Balick 1995). Some of these non-timber forest products are used for traditional medicines and are essential for certain remote local communities with little access to, or desire for, western medicines (Mendelsohn & Balick 1995). Researchers have explored whether it is possible for the local people to extract rattan (*Calamus exilis*) from the Kerinci-Seblat National Park (Sumatra) (Belsky & Siebert 1995). Rattan is a coppicing cane used in local handicrafts and basketry. If rattan extraction can provide local communities with a viable means of earning their livelihood, then there would be less incentive for them to convert the park to agriculture. Belsky & Siebert (1995) proposed that rattan can be extracted sustainably at four year intervals from designated areas of the park. Examples such as rattan extraction provide good illustrations of the practicability of coexistence of protected areas and local communities, and such should be considered as feasible real-world options for shared land use across many parts of Southeast Asia.

In unprotected areas, impacts of human development on native biodiversity could be minimised by discouraging the large-scale and often homogeneous conversion of land for agriculture. For example, some agricultural subsidies could be removed and replaced by other, more environmentally friendly incentives (e.g. alternative employment opportunities) to compensate the impacted local communities (James *et al.* 1999; O'Brien & Kinnaird 2003). Funds saved through elimination of perverse subsidies can be channelled instead towards better biodiversity conservation (James *et al.* 2000). Furthermore, appropriate land use decisions (e.g. retaining sufficiently large and connected parcels of native vegetation) in degraded habitats (e.g. countryside currently partially converted for crops) could substantially increase their long-term conservation value (Sodhi *et al.* 2005c). James *et al.* (1999) also recommended that countries which have signed the Convention on Biological Diversity should make efforts to reduce environmentally damaging subsidies and reroute these resources towards more effective biodiversity conservation.

Let the biology bloom

Large-scale, coordinated research efforts are needed urgently to remedy the paucity of biodiversity studies in Southeast Asia (Sodhi & Liow 2000). This problem is exacerbated by the lack of regional and international expertise on Southeast Asia's rich and unique biodiversity. Conservation efforts, including both strategic and on-ground management, will undoubtedly benefit from a better understanding of the biology of native biotas. A comprehensive programme of environmental and biodiversity mapping must be conducted to reverse the sad state of affairs now confronting Southeast Asian biodiversity. More and better directed research will also help train the local communities, enabling them to better protect the natural resources upon which their livelihood and health depends.

When one considers the diversity of habitat types in Southeast Asia's natural areas (e.g. rain forest, mangroves and caves) (Whitmore 1997), comparatively little ground has been covered on the effects of habitat disturbance on biota, especially over the longer-term. Data are also presently insufficient to evaluate such questions as whether different types of disturbance (e.g. agriculture and urbanisation) provoke similar negative effects on the biota. Some tropical areas remain poorly surveyed and the status of their biodiversity uncertain (Whitten *et al.* 1987a). Detailed biological knowledge is usually needed for proper and effective conservation management (Reed 1999), but such understanding may be deficient for certain tropical areas, such as Southeast Asia (Diamond *et al.* 1987; Sodhi & Liow 2000). Systematic surveys to determine faunal vulnerability and to collect relevant ecological data are needed throughout Southeast Asia in order to practice educated conservation. Life history variables (e.g. reproductive output, dispersal abilities) and inter-specific interactions remain unknown for the large proportion of the taxa. These lacunae makes any present generalisation of extinction-proneness of Southeast Asian biotas ecologically rather shallow.

Studies determining the effects and consequences of habitat disturbance on Southeast Asian biotas must be better designed (Bierregaard *et al.* 1997; Crome 1997; Danielsen 1997; Sodhi *et al.* 2004a). These studies must attempt to: 1) obtain adequate sample sizes, 2) have adequate sampling periods with multiple visits, 3) use more than one sampling method (e.g. point counts and mist-netting) for satisfactory sampling, 4) estimate species abundances or densities, preferably over time and space, 5) validate local extinctions using playback surveys and present species accumulation curves (Gotelli & Colwell 2001), 6) have adequate controls (e.g. larger 'original' forests) and better replication, where possible, and 7) seek a better understanding of the functional attributes of key species – crucial for assessing the resilience of ecosystem function in the face of large-scale disturbance (du Toit *et al.* 2004). Just as importantly, data sets from such studies should be both published in the peer-reviewed scientific literature and strident efforts should be made to promote a wider access (e.g. via the internet and mass media) of the key findings and conclusions to non-scientific clientele, resource managers and the general public. We also recommend that individuals making inventories rank species according to their abundance and clearly define their ranking system. This will assist in determining whether a species is declining in abundance (which can provide political and social leverage for 'State of the Environment' audits) and whether rare species are more extinction-prone than common species.

Scientists must collect more data showing whether applied conservation measures (e.g. establishment of reserves and adoption of alternative land use methods) in Southeast Asia are effective in abating extinctions and arresting population declines. More information is also needed about the value of biodiversity in maintaining vital ecological processes, in addition to providing ecosystem services (pollination of crops) to humanity (see Chapter 1). Bawa *et al.* (2004a) detail priorities for conservation research in the tropics.

Summary

1. Massive anthropogenically driven habitat modifications, forest fires, and the overexploitation of wildlife in Southeast Asia, are the most clear-and-present dangers to the region's biodiversity (see Chapters 3 and 4).

2. In spite of the generally pessimistic outlook, there are feasible ways to conserve at least some of the regional natural resources.
3. Key solutions should include enhancing public environmental awareness (which will also help to change political decision-making), delineating adequately protected reserves and providing economic incentives for conservation.
4. Good governance, political accountability, independent and fair judiciary, and freer press all facilitate biodiversity conservation and should be encouraged.
5. Given that many of the drivers of biodiversity loss (e.g. international demand for rain forest timber; global warming associated with elevation in CO_2 levels) are issues that transcend national boundaries, any realistic solution will need to involve a multinational and multi-disciplinary strategy, including political, socio-economic and scientific input, in which all major stakeholders (governmental, non-governmental, national and international organizations) must partake.

REFERENCES

Achard, F., Eva, H. D., Stibig, H.-J. *et al.* (2002). Determination of deforestation rates of the world's humid tropical forests. *Science*, **297**, 999–1002.

Adams, W. M., Aveling, R., Brockington, D. *et al.* (2004). Biodiversity conservation and the eradication of poverty. *Science*, **306**, 1146–9.

Adeel, Z. & Pomeroy, R. (2002). Assessment and management of mangrove ecosystems in developing countries. *Trees*, **16**, 235–8.

Aiken, S. R. & Leigh, C. H. (1992). *Vanishing Rain Forests.* Oxford: Clarendon Press.

Akbar, Z. & Ariffin, M. K. E. (1997). A comparison of small mammal abundance between a primary and disturbed lowland rain forest in Peninsular Malaysia. *Malayan Nature Journal*, **50**, 201–6.

Aldous, P. (2004). Borneo is burning. *Nature*, **432**, 144–6.

Alvard, M. S. & Winarni, N. L. (1999). Avian biodiversity in Morowali Nature Reserve, Central Sulawesi, Indonesia and the impact of human subsistence activities. *Tropical Biodiversity*, **6**, 59–74.

An, Z. (2000). The history and variability of the East Asian paleomonsoon climate. *Quaternary Science Reviews*, **19**, 171–87.

Anonymous (2004a). *We've 'Proof' of Wood-smuggling: Greenpeace.* The Straits Times (7th October 2004).

Anonymous (2004b). *Graft and Poverty at Root of Deforestation.* The Straits Times (4th December 2004).

Anonymous (2004c). *Fresh Tiger Meat on the Menu.* The Straits Times (7th December 2004).

Aplet, G. H., Anderson, S. J. & Stone, C. P. (1991). Association between feral pig disturbance and the composition of some alien plant assemblages in Hawaii Volcanoes National Park. *Vegetatio*, **95**, 55–62.

Appanah, S. (1993). Mass flowering of dipterocarp forests in the aseasonal tropics. *Journal of Biosciences*, **18**, 457–74.

Appanah, S. (1998). Management of natural forests. In *A Review of Dipterocarps, Taxonomy, Ecology and Silviculture*, ed. S. Appanah & J. M. Turnbull. Bogor Barat, Indonesia: CIFOR, pp. 133–49.

Argeloo, M. & Dekker, R. W. R. J. (1996). Exploitation of megapode eggs in Indonesia: the role of traditional methods in the conservation of megapodes. *Oryx*, **30**, 59–64.

Asia–Pacific Migratory Waterbird Conservation Committee (2001). *Asia–Pacific Waterbird Conservation Strategy: 2001–05.* Kuala Lumpur, Malaysia: Wetlands International-Asia Pacific.

Audley-Charles, M. G. (1983). Reconstruction of eastern Gondwanaland. *Nature*, **306**, 48–50.

Augspurger, C. K. (1981). Reproductive synchrony of a tropical shrub: experimental studies on effects of pollinators and seed predators on *Hybanthus prunifolius* (Violaceae). *Ecology*, **62**, 775–88.

Balmford, A. (1996). Extinction filters and current resilience: the significance of past selection pressures for conservation biology. *Trends in Ecology & Evolution*, **11**, 193–6.

Balmford, A. & Whitten, T. (2003). Who should pay for tropical conservation, and how could the costs be met? *Oryx*, **37**, 238–50.

Balmford, A., Bruner, A., Cooper, P. *et al.* (2002). Economic reasons for conserving wild nature. *Science*, **297**, 950–3.

Balmford, A., Green, R. E. & Jenkins, M. (2003a). Measuring the changing state of nature. *Trends in Ecology & Evolution*, **18**, 326–30.

Balmford, A., Gaston, K. J., Blyth, S., James, A. & Kapos, V. (2003b). Global variation in terrestrial conservation costs, conservation benefits, and unmet conservation needs. *Proceedings of the National Academy of Sciences of the United States of America*, **100**, 1046–50.

Baran, E. & Hambrey, J. (1998). Mangrove conservation and coastal mangement in Southeast Asia: what impact on fishery resources? *Marine Pollution Bulletin*, **37**, 431–40.

Barbier, E. B. (1993). Economic aspects of tropical deforestation in Southeast Asia. *Global Ecology and Biogeography Letters*, **3**, 215–34.

Barr, C. (2001). Banking on sustainability: structural adjustment and forestry reform in post-Suharto Indonesia. Washington, DC: WWF Macroeconomics Program Office and CIFOR.

Bawa, K. S. (1990). Plant-pollinator interactions in tropical rain forests. *Annual Review of Ecology and Systematics*, **21**, 399–422.

Bawa, K. S. & Dayanandan, S. (1997). Socioeconomic factors and tropical deforestation. *Nature*, **386**, 562–3.

Bawa, K. S., Kress, W. J. & Nadkarni, N. M. (2004a). Beyond paradise – meeting the challenges in tropical biology in the 21st century. *Biotropica*, **36**, 276–84.

Bawa, K. S., Kress, W. J., Nadkarni, N. M. *et al.* (2004b). Tropical ecosystems into the 21st century. *Science*, **206**, 227–8.

Beck, J., Schulze, C. H., Linsenmair, K. E. & Fiedler, K. (2002). From forest to farmland: diversity of geometrid moths along two habitat gradients on Borneo. *Journal of Tropical Ecology*, **18**, 33–51.

Beebee, T. J. C. (1992). Amphibian decline. *Nature*, **355**, 120.

Beissinger, S. R. (2001). Trade of live birds: potential, principles and practices of sustainable use. In *Conservation of Exploited Species*, ed. J. D. Reynolds, G. M. Mace & J. G. Robinson. Cambridge: Cambridge University Press, pp. 183–202.

Belsky, J. M. & Siebert, S. F. (1995). Managing rattan harvesting for local livelihoods and forest conservation in Kerinci-Seblat National Park, Sumatra. *Selbyana*, **16**, 212–22.

Bennett, E. L. (2000). Timber certification: where is the voice of the biologist? *Conservation Biology*, **14**, 921–33.

Bennett, E. L. (2002). Is there a link between wild meat and food security? *Conservation Biology*, **16**, 590–2.

Bennett, E. L. & Caldecott, J. O. (1981). Unexpected abundance: the trees and wildlife of the Lima Belas Estate forest reserve, near Slim River, Perak. *The Planter*, **57**, 516–19.

Bennett, E. L., Nyaoi, A. J. & Sompud, J. (2000). Saving Borneo's bacon: the sustainability of hunting in Sarawak and Sabah. In *Hunting for Sustainability in Tropical Forests*, ed. J. G. Robinson & E. L. Bennett. New York: Columbia University Press, pp. 305–24.

Berbet, M. L. C. & Costa, M. H. (2003). Climate change after tropical deforestation: seasonal variability of surface albedo and its effects on precipitation change. *Journal of Climate*, **16**, 2099–104.

Berkes, F. (2004). Rethinking community-based conservation. *Conservation Biology*, **18**, 621–30.

Bertault, J.-G. & Sist, P. (1997). An experimental comparison of different harvesting intensities with reduced-impact and conventional logging in East Kalimantan, Indonesia. *Forest Ecology and Management*, **94**, 209–18.

Bierregaard, R. O. J., Laurance, W. F., Site, J. W. J., Lynam, A. J. & Didham, R. K. (1997). Key priorities for the study of fragmented tropical ecosystems. In *Tropical Forest Remnants: Ecology, Management, and Conservation of Fragmented Communities*, ed. W. F. Laurance & J. R. O. Bierregaard. Chicago: University of Chicago Press, pp. 515–25.

Billington, C., Kapos, V., Edwards, M., Blyth, S. & Iremonger, S. (1996). *Estimated Original Forest Cover Map – A First Attempt*. Cambridge, UK: WCMC.

Blaustein, A. R. & Kiesecker, J. M. (2002). Complexity in conservation: lessons from the global decline of amphibian populations. *Ecology Letters*, **5**, 597–608.

Blower, J., Paine, J., Hahn, U. S., Ohn, U. & Sutter, H. (1991). Burma (Myanmar). In *The Conservation Atlas of Tropical Forests: Asia and the Pacific*, ed. N. M. Collins, J. A. Sayer & T. C. Whitmore. London: Macmillan Press Ltd, pp. 103–110.

Bolle, H.-J., Seiler, W. & Bolin, B. (1986). *The Greenhouse Effect, Climate Change and Ecosystems*. New York: Wiley and Sons.

Bourliére, F. (1973). The comparative ecology of rain forest mammals in Africa and tropical America: some introductory remarks. In *Tropical Forest Ecosystems in Africa and South America: A Comparative Review*, ed. B. J. Meggers, E. S. Ayensu & W. D. Duckworth. Washington, DC: Smithsonian Institution Press, pp. 279–92.

Bowles, I. A., Rice, R. E., Mittermeier, R. A. & da Fonseca, G. A. B. (1998). Logging and tropical forest conservation. *Science*, **280**, 1899–900.

Bowman, D. M. J. S. (2002). The Australian summer monsoon: a biogeographic perspective. *Australian Geographical Studies*, **40**, 261–77.

Brandon-Jones, D. (1998). Pre-glacial Bornean primate impoverishment and Wallace's Line. In *Biogeography and Geological Evolution of SE Asia*, ed. R. Hall & J. D. Holloway. Leiden: Backhuys, pp. 393–404.

Brook, B. W. & Bowman, D. (2004). The uncertain blitzkrieg of Pleistocene megafauna. *Journal of Biogeography*, **31**, 517–23.

Brook, B. W. & Whitehead, P. J. (2004). Sustainable harvest regimes for magpie geese (*Anseranas semipalmata*) under spatial and temporal heterogeneity. *Wildlife Research*, in press.

Brook, B. W., Sodhi, N. S. & Ng, P. K. L. (2003). Catastrophic extinctions follow deforestation in Singapore. *Nature*, **424**, 420–3.

Brooks, T. M., Pimm, S. L. & Collar, N. J. (1997). Deforestation predicts the number of threatened birds in insular Southeast Asia. *Conservation Biology*, **11**, 382–94.

Brooks, T. M., Pimm, S. L., Kapos, V. & Ravilious, C. (1999). Threat from deforestation to montane and lowland birds and mammals in insular South-east Asia. *Journal of Animal Ecology*, **68**, 1061–78.

Brown, N., Press, M. & Bebber, D. (1999). Growth and survivorship of dipterocarp seedlings: differences in shade persistence create a special case of dispersal limitation. *Philosophical Transactions of the Royal Society of London B Biological Sciences*, **54**, 1847–55.

Bruhl, C. A., Eltz, T. & Linsenmair, K. E. (2003). Size does matter: effects of tropical rainforest fragmentation on the leaf litter ant community in Sabah, Malaysia. *Biodiversity and Conservation*, **12**, 1371–89.

Bruner, A. G., Gullison, R. E., Rice, R. E. & da Fonseca, G. A. B. (2001). Effectiveness of parks in protecting tropical biodiversity. *Science*, **291**, 125–8.

Bryant, R. L. (2002). False prophets? Mutant NGOs and Philippine environmentalism. *Society and Natural Resources*, **15**, 629–39.

Bryant, R. L., Rigg, J. & Stott, P. (1993). Forest transformations and political ecology in Southeast-Asia. *Global Ecology and Biogeography Letters*, **3**, 101–11.

Buffetaut, E. & Suteethorn, V. (1998). The biogeographical significance of the Mesozoic vertebrates from Thailand. In *Biogeography and Geological Evolution of SE Asia*, ed. R. Hall & J. D. Holloway. Leiden: Backhuys, pp. 83–90.

Buij, R., Wich, S. A., Lubis, A. H. & Sterck, E. H. M. (2002). Seasonal movements in the Sumatran orangutan (*Pongo pygmaeus abelii*) and consequences for conservation. *Biological Conservation*, **107**, 83–7.

Bulte, E. H. & Horan, R. D. (2002). Does human population growth increase wildlife harvesting? An economic assessment. *Journal of Wildlife Management*, **66**, 574–80.

Bush, M. B. (1994). Amazonian speciation: a necessarily complex model. *Journal of Biogeography*, **21**, 5–17.

Butchart, S. H. M. & Baker, G. C. (2000). Priority sites for conservation of maleos (*Macrocephlon maleo*) in central Sulawesi. *Biological Conservation*, **94**, 79–91.

Butlin, R. K., Walton, C., Monk, K. A. & Bridle, J. R. (1998). Biogeography of Sulawesi grasshoppers, genus *Chitaura*, using DNA sequence data. In *Biogeography and Geological Evolution of SE Asia*, ed. R. Hall & J. D. Holloway. Leiden: Backhuys, pp. 355–9.

Byron, N. & Waugh, G. (1988). Forestry and fisheries in Asian–Pacific Region: issues in natural resource management. *Asian–Pacific Economic Literature*, **2**, 46–80.

Cahill, A. J. & Walker, J. S. (2000). The effects of forest fire on the nesting success of the red-knobbed hornbill *Aceros cassidix*. *Bird Conservation International*, **10**, 109–14.

Caldecott, J. O., Blouch, R. A. & Macdonald, A. A. (1993). The Bearded Pig (*Sus barbatus*) *Pigs, Peccaries and Hippos Status Survey and Action Plan*. Gland, Switzerland: IUCN/SSC.

Caniago, I. & Siebert, S. F. (1998). Medicinal plant ecology, knowledge and conservation in Kalimantan, Indonesia. *Economic Botany*, **52**, 229–50.

Cannon, C. H., Peart, D. R., Leighton, M. & Kartawinata, K. (1994). The structure of lowland rain forest after selective logging in West Kalimantan, Indonesia. *Forest Ecology and Management*, **67**, 49–68.

Cannon, C. H., Peart, D. R. & Leighton, M. (1998). Tree species diversity in commercially logged Bornean rainforest. *Science*, **281**, 1366–7.

Cardillo, M., Purvis, A., Sechrest, W. *et al.* (2004). Human population density and extinction risk in the world's carnivores. *PLoS Biology*, **2**, 0909–14.

Casellini, N., Foster, K. & Hien, B. T. T. (1999). *The 'White Gold' of the Sea: A Case Study of Sustainable Harvesting of Swiftlet Nest in Coastal Vietnam*. Gland, Switzerland: IUCN.

Cassola, F. (1996). Studies on the tiger beetles: LXXXIV. Additions to the tiger beetle fauna of Sulawesi, Indonesia (Coleoptera: Cicindelidae). *Zoologische Medelingen (Leiden)*, **70**, 145–53.

Castelletta, M., Sodhi, N. S. & Subaraj, R. (2000). Heavy extinctions of forest avifauna in Singapore: lessons for biodiversity conservation in Southeast Asia. *Conservation Biology*, **14**, 1870–80.

Castelletta, M., Thiollay, J.-M. & Sodhi, N. S. (2005). The effects of extreme forest fragmentation on the bird community of Singapore island. *Biological Conservation*, **121**, 135–55.

Caughley, G. & Gunn, A. (1986). *Conservation Biology in Theory and Practice*. Cambridge, MA: Blackwell Science.

Center for International Forestry Research [CIFOR] (2003). *CIFOR Annual Report 2003: Science for Forests and People*. Bogor Barat, Indonesia: CIFOR.

Chaimanee, Y. (1999). Plio-Pleistocene rodents of Thailand. *Thai Studies in Biodiversity*, **3**, 1–303.

Chape, S., Fish, L., Fox, P. & Spalding, M. (2003). *United Nations List of Protected Areas*. Gland, Switzerland/Cambridge, UK: IUCN/UNEP.

Chey, V. K. (2000). Moth diversity in the tropical rain forest of Lanjak-Entimau, Sarawak, Malaysia. *Malayan Nature Journal*, **54**, 305–18.

Chivian, E. (2002). *Biodiversity: Its Importance to Human Health*. Cambridge, MA: Center for Health and the Global Environment, Harvard Medical School.

Christiansen, M. B. & Pitter, E. (1997). Species loss in a forest birds community near Lagoa Santa in Southern Brazil. *Biological Conservation*, **80**, 23–32.

Christy, M. (2002). Sulawesi's disappearing flagship bird. *Species*, **38**, 8–9.

Chua, K. B., Bellini, W. J., Rota, P. A. *et al.* (2000). Nipah virus: a recently emergent deadly paramyxovirus. *Science*, **288**, 1432–5.

Chung, F. J. (1996). Interests and policies of the state of Sarawak, Malaysia regarding intellectual property rights for plant derived drugs. *Journal of Ethnopharmacology*, **51**, 201–4.

Ciofi, C., Beaumont, M. A., Swingland, I. R. & Bruford, M. W. (1999). Genetic divergence and units for conservation in the Komodo dragon *Varanus komodoenis*. *Proceedings of the Royal Society of London Series B Biological Sciences*, **266**, 2269–74.

Clayton, D. H. & Milner-Gulland, E. J. (2000). The trade in wildlife in north Sulawesi, Indonesia. In *Hunting for Sustainability in Tropical Forests*, ed. J. G. Robinson & E. L. Bennett. New York: Columbia University Press, pp. 473–98.

Clayton, D. H., Keeling, M. & Milner-Gulland, E. J. (1997). Bringing home the bacon: a spatial model of wild pig hunting in Sulawesi, Indonesia. *Ecological Applications*, **7**, 642–52.

Cleary, D. F. R. (2003). An examination of scale of assessment, logging and ENSO-induced fires on butterfly diversity in Borneo. *Oecologia*, **135**, 313–21.

Cleary, D. F. R. & Genner, M. J. (2004). Changes in rain forest butterfly diversity following major ENSO-induced fires in Borneo. *Global Ecology and Biogeography*, **13**, 129–40.

Coates, B. J. & Bishop, K. D. (1997). *A Guide to the Birds of Wallacea*. Alderley, Queensland: Dove Publications.

Cochrane, M. A. (2003). Fire science for rainforests. *Nature*, **421**, 913–19.

Coley, P. D. (1998). Possible effects of climate change on plant/herbivore interactions in moist tropical forests. *Climate Change*, **39**, 455–72.

Collins, J. P. & Storfer, A. (2003). Global amphibian declines: sorting the hypotheses. *Diversity and Distributions*, **9**, 89–98.

Collins, M. (1992). Introduction. In *The Conservation Atlas of Tropical Forests: Africa*, ed. J. A. Sayer, C. S. Harcourt & N. M. Collins. New York: IUCN/Macmillan, pp. 9–16.

Collins, N. M. & Morris, M. G. (1985). *Threatened Swallowtail Butterflies of the World. The IUCN Red Data Book*. Gland, Switzerland/Cambridge, UK: IUCN.

Colón, C. P. (2002). Ranging behaviour and activity of the Malay civet (*Viverra tangalunga*) in a logged and an unlogged forest in Danum Valley, east Malaysia. *Journal of Zoology*, **257**, 473–85.

Cooper, D. S. & Francis, C. M. (1998). Nest predation in a Malaysian lowland rain forest. *Biological Conservation*, **85**, 199–202.

Corbet, A. S. & Pendlebury, H. M. (1992). *The Butterflies of the Malay Peninsula*, 4th edn. Kuala Lumpur: Malay Nature Society.

Corbet, G. B. & Hill, J. E. (1992). *The Mammals of the Indomalayan Region: a Systematic Review*. Oxford: Oxford University Press.

Corlett, R. T. (1988). The naturalized flora of Singapore. *Journal of Biogeography*, **15**, 657–63.

Corlett, R. T. (1991). Vegetation. In *The Biophysical Environment of Singapore*, ed. L. S. Chia, A. Rahman & D. B. H. Tay. Singapore: Singapore University Press, pp. 134–54.

Corlett, R. T. (1992). The ecological transformation of Singapore 1819–1990. *Journal of Biogeography*, **19**, 411–20.

Corlett, R. T. (2000). Environmental heterogeneity and species survival in degraded tropical landscapes. In *The Ecological Consequences of Environmental Heterogeneity*, ed. M. J. Hutchings, E. A. John & A. Stewart. Oxford: Blackwell Science, pp. 333–55.

Corlett, R. T. (2005). Vegetation. In *The Physical Geography of Southeast Asia*, ed. A. Gupta. Oxford: Oxford University Press, pp. 105–19.

Corrigan, P. J. (1992). *Investigation of the Southern Thailand Zebra Dove Industry*. *TRAFFIC Southeast Asia Field Report No. 1*. Selangor, Malaysia.

Cox, R. (1991). Philippines. In *The Conservation Atlas of Tropical Forests: Asia and the Pacific*, ed. N. M. Collins, J. A. Sayer & T. C. Whitmore. New York: IUCN/Simon & Schuster, pp. 192–200.

Crane, E. & Walker, P. (1983). *The Impact of Pest Management on Bees and Pollination*. London, UK: Tropical Development and Research Institute.

Crome, F. H. J. (1997). Researching tropical forest fragmentation: shall we keep doing what we're doing? In *Tropical Forest Remnants: Ecology, Management, and Conservation of Fragmented Communities*, ed. W. F. Laurance & R. O. Bierregaard. Chicago: University of Chicago Press, pp. 485–501.

Curran, L. M. & Leighton, M. (2000). Vertebrate responses to spatiotemporal variation in seed production of mast-fruiting Dipterocarpaceae. *Ecological Monographs*, **70**, 101–28.

Curran, L. M. & Webb, C. O. (2000). Experimental tests of the spatiotemporal scale of seed predation in mast-fruiting Dipterocarpaceae. *Ecological Monographs*, **70**, 129–48.

Curran, L. M., Caniago, I., Paoli, G. D. *et al.* (1999). Impact of El Niño and logging on canopy tree recruitment in Borneo. *Science*, **286**, 2184–8.

Curran, L. M., Trigg, S. N., McDonald, A. K. *et al.* (2004). Lowland forest loss in protected areas of Indonesian Borneo. *Science*, **303**, 1000–3.

Currie, D. (1991). Energy and large-scale patterns of animal- and plant-species richness. *American Naturalist*, **137**, 227–49.

Daily, G. C. & Walker, B. H. (2000). Seeking the great transition. *Nature*, **403**, 243–5.

Danielsen, F. (1997). Stable environments and fragile communities: does history determine the resilience of avian rain-forest communities to habitat degradation? *Biodiversity and Conservation*, **6**, 423–33.

Danielsen, F. & Treadaway, C. G. (2004). Priority conservation areas for butterflies (Lepidoptera: Rhopalocera) in the Philippine islands. *Animal Conservation*, **7**, 79–92.

Davies, R. G., Eggleton, P., Jones, D. T., Gathorne-Hardy, F. J. & Hernandez, L. M. (2003). Evolution of termite functional diversity: analysis and synthesis of local ecological and regional influences on local species richness. *Journal of Biogeography*, **30**, 847–77.

Davis, A. J., Holloway, J. D., Huijbregts, H. *et al.* (2001). Dung beetles as indicators of change in the forests of northern Borneo. *Journal of Applied Ecology*, **38**, 593–616.

Davis, G. (2002). Bushmeat and international development. *Conservation Biology*, **16**, 587–9.

DeFries, R. S., Houghton, R. A., Hansen, M. C. *et al.* (2002). Carbon emissions from tropical deforestation and regrowth based on satellite observations for the 1980s and 1990s. *Proceedings of the National Academy of Sciences of the United States of America*, **99**, 14256–61.

de Jong, R. (1998). Halmahera and Seram: different histories, but similar butterfly faunas. In *Biogeography and Geological Evolution of SE Asia*, ed. R. Hall & J. D. Holloway. Leiden: Backhuys, pp. 315–25.

Dekker, R. W. R. J. (1990). The distribution and status of nesting grounds of the maleo *Macrocephalon maleo* in Sulawesi, Indonesia. *Biological Conservation*, **51**, 139–50.

Diamond, J. M. (1980). Patchy distributions of tropical birds. In *Conservation Biology: An Evolutionary–Ecological Perspective*, ed.

M. E. Soulé & B. A. Wilcox. New York: Sinauer, pp. 57–74.

Diamond, J. M., Bishop, K. D. & van Balen, S. (1987). Bird survival in an isolated Javan Indonesia woodland island or mirror. *Conservation Biology*, **1**, 132–42.

Director General of Forest Protection and Nature Conservation [PHPA] (1998). *Species Recovery Plan Yellow-crested Cockatoo*. Bogor, Indonesia: PHPA/LIPI/Birdlife International-IP.

Dirzo, R. & Raven, P. J. (2003). Global state of biodiversity and loss. *Annual Review of Environment and Resources*, **28**, 137–67.

Dodson, J. J., Colombani, F. & Ng, P. K. L. (1995). Phylogeographic structure in mitochondrial DNA of a South-east Asian freshwater fish, *Hemibagrus nemurus* (Siluroidea; Bagridae) and Pleistocene sea-level changes on the Sunda shelf. *Molecular Ecology*, **4**, 331–46.

Durst, P. B., Killman, W. & Brown, C. (2004). Asia's new woods. *Journal of Forestry*, June, 47–53.

du Toit, J. T., Walker, B. H. & Campbell, B. M. (2004). Conserving tropical nature: current challenges for ecologists. *Trends in Ecology & Evolution*, **19**, 12–17.

Eames, J. C., Trai, L. T. & Cu, N. (1999a). A new species of laughingthrush (Passeriformes: Garrulacinae) from the Western Highlands of Vietnam. *Bulletin British Ornithological Club*, **119**, 4–15.

Eames, J. C., Trai, L. T. Cu, N. & Eve, R. (1999b). New species of barwing *Actinodura* (Passeriformes: Sylviinae: Timaliini) from the Western Highlands of Vietnam. *Ibis*, **141**, 1–10.

Eggleton, P., Bignell, D. E., Sands, W. A. *et al.* (1995). The species richness of termites (Isoptera) under differing levels of forest disturbance in the Mbalmayo Forest Reserve, southern Cameroon. *Journal of Tropical Ecology*, **11**, 85–98.

Eggleton, P., Homathevi, R., Jeeva, D. *et al.* (1997). The species richness and composition of termites (Isoptera) in primary and regenerating lowland dipterocarp forest in Sabah, East Malaysia. *Ecotropica*, **3**, 119–28.

Eggleton, P., Homathevi, R., Jones, D. T. *et al.* (1999). Termite assemblages, forest disturbance and greenhouse gas fluxes in Sabah, East Malaysia. *Philosophical Transactions of the Royal Society of London B Biological Sciences*, **354**, 1791–802.

Ehrlich, P. & Ehrlich, A. (1996). *Betrayal of Science and Reason: How Anti-environmental Rhetoric Threatens our Future*. Washington, DC: Island Press.

Eltz, T., Bruhl, C. A., van der Kaars, S. & Linsenmair, K. E. (2002). Determinants of stingless bee nest density in lowland dipterocarp forests of Sabah, Malaysia. *Oecologia*, **131**, 27–34.

Eltz, T., Bruehl, C. A., Imiyabir, Z. & Linsenmair, K. E. (2003). Nesting and nest trees of stingless bees (Apidae: Meliponini) in lowland dipterocarp forests in Sabah, Malaysia, with implications for forest management. *Forest Ecology and Management*, **172**, 301–13.

Er, K. B. H., Vardon, M. J., Tanton, M. T., Tidermann, C. R. & Webb, G. J. W. (1995). *Edible Birds' Nest Swiftlets and CITES: a Review of the Evidence of Population Decline and Nest Harvesting Effects*. Centre for Resource and Environmental Studies Working Paper 1995/3. Canberra: Australian National University.

Eudey, A. A. (1994). Temple and pet primates in Thailand. *Revue D Ecologie-la Terre Et La Vie*, **49**, 273–80.

Fa, J. E., Juste, J., Burn, R. W. & Broad, G. (2002). Bushmeat consumption and preferences of two ethnic groups in Bioko Island, west Africa. *Human Ecology*, **30**, 397–416.

Fearnside, P. M. & Laurance, W. F. (2003). Comment on 'Determination of deforestation rates of the world's humid tropical forests'. *Science*, **299**, 1015a.

Felton, A. M., Engstrom, L. M., Felton, A. & Knott, C. D. (2003). Orangutan population density, forest structure and fruit availability in hand-logged and unlogged peat swamp forests in west Kalimantan, Indonesia. *Biological Conservation*, **114**, 91–101.

Ferguson, N. M., Fraser, C., Donnelly, C. A., Ghani, A. C. & Anderson, R. M. (2004). Public health risk from the avian H5N1 influenza epidemic. *Science*, **304**, 968–9.

Fermon, H., Waltert, M., Vane-Wright, R. I. & Muhlenberg, M. (2003). Forest use and vertical stratification in fruit-feeding butterflies of Sulawesi, Indonesia: impacts for conservation. *Biodiversity and Conservation*, **12**, 1–18.

Fernandez-Duque, E. & Valeggia, C. (1994). Meta-analysis: a valuable tool in conservation research. *Conservation Biology*, **8**, 555–61.

Flannery, T. F. (1996). Mammalian zoogeography of New Guinea and the southwestern Pacific. In *The Origin and Evolution of Pacific Island biotas, New Guinea to Eastern Polynesia: Patterns and Processes*, ed. A. Keast & S. E. Miller. Amsterdam: SPB Academic Publishing, pp. 399–406.

Flint, E. P. (1994). Changes in land use in South and Southeast Asia from 1880 to 1980: a data base prepared as part of a coordinated research program on carbon fluxes in the tropics. *Chemosphere*, **29**, 1015–62.

Floren, A. & Linsenmair, K. E. (2001). The influence of anthropogenic disturbances on the structure of arboreal arthropod communities. *Plant Ecology*, **153**, 153–67.

Floren, A., Freking, A., Biehl, M. & Linsenmair, K. E. (2001). Anthropogenic disturbance changes the structure of arboreal tropical ant communities. *Ecography*, **24**, 547–54.

Food and Agriculture Organization [FAO] (1993). *Forest Resources Assessment 1990: Tropical Countries. FAO Forestry Paper 112.* Rome: United Nations Food and Agriculture Organization.

Food and Agriculture Organization [FAO] (2004). *The State of Food and Agriculture 2003–2004.* Rome: United Nations Food and Agriculture Organization.

Foody, G. M. & Cutler, M. E. J. (2003). Tree biodiversity in protected and logged Bornean tropical rain forests and its measurement by satellite remote sensing. *Journal of Biogeography*, **30**, 1053–66.

Forest Resources Assessment [FRA] (2000). *Forest Resources Assessment 1990: Global Synthesis. FAO Forestry Paper 124.* Rome: United Nations Food and Agriculture Organization.

Fortey, R. A. & Cocks, L. R. M. (1998). Biogeography and palaeogeography of the Sibumasu terrrane in the Ordovician: a review. In *Biogeography and Geological Evolution of SE Asia*, ed. R. Hall & J. D. Holloway. Leiden: Backhuys, pp. 43–56.

Francis, C. (2001). *Mammals of South-East Asia.* London: New Holland.

Frankham, R., Ballou, J. D. & Briscoe, D. A. (2002). *Introduction to Conservation Genetics.* Cambridge: Cambridge University Press.

Gaston, K. J. & Spicer, J. I. (1998). *Biodiversity: an Introduction.* Oxford: Blackwell Science.

Gathorne-Hardy, F. J., Jones, D. T. & Syaukani, D. R. G. (2002a). A regional perspective on the effects of human disturbance on the termites of Sundaland. *Biodiversity and Conservation*, **11**, 1991–2006.

Gathorne-Hardy, F. J., Syaukani, D. R. G., Eggleton, P. & Jones, D. T. (2002b). Quaternary rainforest refugia in south-east Asia: using termites (Isoptera) as indicators. *Biological Journal of the Linnean Society*, **75**, 453–66.

Geist, H. J. & Lambin, E. F. (2002). Proximate causes and underlying driving forces of tropical deforestation. *Bioscience*, **52**, 143–50.

Ghazoul, J. (2002). Impact of logging on the richness and diversity of forest butterflies in a tropical dry forest in Thailand. *Biodiversity and Conservation*, **11**, 521–41.

Ghosh, N. (2004). *Avian Flu Virus found in Thai Wild Birds.* The Straits Times (15th December).

Giri, C., Defourny, P. & Shrestha, S. (2003). Land cover characterization and mapping of continental Southeast Asia using multiresolution satellite sensor data. *International Journal of Remote Sensing*, **24**, 4181–96.

Goerck, J. M. (1997). Patterns of rarity in the birds of the Atlantic forest of Brazil. *Conservation Biology*, **11**, 112–18.

Good, R. (1974). *The Geography of the Flowering Plants*, 4th edn. London: Longman.

Gorog, A. J., Sinaga, M. H. & Engstrom, M. D. (2004). Vicariance or dispersal? Historical biogeography of three Sunda shelf murine rodents (*Maxomys surifer, Leopaldamys sabanus* and *Maxomys whiteheadi*). *Biological Journal of the Linnean Society*, **81**, 91–109.

Gotelli, N. J. & Colwell, R. K. (2001). Quantifying biodiversity: procedures and pitfalls in the measurement and comparison of species richness. *Ecology Letters*, **4**, 379–91.

Grabherr, G., Gottfried, M. & Pauli, H. (1994). Climate effects on mountain plants. *Nature*, **369**, 448.

Groombridge, B. & Luxmoore, R. (1991). *Pythons in South-east Asia. A review of Distribution, Status and Trade in Three Selected Species.* Convention on International Trade in Endangered Species (CITES) Secretariat, Laussane, Switzerland.

Grove, R. H., Damodaran, V. & Sangwan, S. (eds.) (1998). *Nature and the Orient: the Environmental History of South and Southeast Asia.* Delhi: Oxford University Press.

Groves, C. P. (2001). *Primate Taxonomy.* Washington, DC: Smithsonian Institution.

Guan, Y., Zheng, B. J., He, Y. Q. *et al.* (2003). Isolation and characterization of viruses related to the SARS Coronavirus from animals in Southern China. *Science*, **302**, 276–8.

Haffer, J. (1969). Speciation in Amazonian birds. *Science*, **165**, 131–7.

Haffer, J. (1974). *Avian Speciation in Tropical South America*. Cambridge, Massachusetts: Nuttal Ornithological Club.

Haffer, J. (1987). Quaternary history of tropical America. In *Biogeography and Quaternary History in Tropical America*, ed. T. C. Whitmore & G. T. Prance. Oxford: Clarendon Press, pp. 1–18.

Hamann, A. & Curio, E. (1999). Interactions among frugivores and fleshy fruit trees in a Philippine submontane rainforest. *Conservation Biology*, **13**, 766–73.

Hamer, K. C., Hill, J. K., Lace, L. A. & Langan, A. M. (1997). Ecological and biogeographical effects of forest disturbance on tropical butterflies of Sumba, Indonesia. *Journal of Biogeography*, **24**, 67–75.

Hamer, K. C., Hill, J. K., Benedick, S. *et al.* (2003). Ecology of butterflies in natural and selectively logged forests of northern Borneo: the importance of habitat heterogeneity. *Journal of Applied Ecology*, **40**, 150–62.

Hamilton, A. C. (1982). *Environmental History of East Africa; A Study of the Quaternary*. London: Academic Press.

Hannah, L., Carr, J. L. & Lankerani, A. (1995). Human disturbance and natural habitat: a biome level analysis of a global data set. *Biodiversity and Conservation*, **4**, 128–55.

Harcourt, C. S., Billington, C. & Sayer, J. A. (1996). Introduction. In *The Conservation Atlas of Tropical Forests: the Americas*, ed. C. S. Harcourt, J. A. Sayer & C. Billington. New York: IUCN/Simon & Schuster, pp. 9–16.

Hardter, R., Woo, Y. C. & Ooi, S. H. (1997). Intensive plantation cropping, a source of sustainable food and energy production in the tropical rain forest areas in southeast Asia. *Forest Ecology and Management*, **93**, 93–102.

Harrison, C. G. A., Brass, G. W., Salzman, E. *et al.* (1981). Sea level variations, global sedimentation rates and the hypsographic curve. *Earth and Planetary Science Letters*, **54**, 1–16.

Harrison, J. L. (1969). The abundance and population density of mammals in Malayan lowland forests. *Malayan Nature Journal*, **22**, 174–8.

Harrison, R. D. (2000). Repercussions of El Niño: drought causes extinction and the breakdown of mutualism in Borneo.

Proceedings of the Royal Society of London Series B Biological Sciences, **267**, 911–15.

Harrison, R. D., Hamid, A. A., Kenta, T. *et al.* (2003). The diversity of hemi-epiphytic figs (*Ficus*; Moraceae) in a Bornean lowland rain forest. *Biological Journal of the Linnean Society*, **78**, 439–55.

Hartley, S. E. & Jones, T. H. (2003). Plant diversity and insect herbivores: effects of environmental changes in contrasting models. *Oikos*, **101**, 6–17.

Hartshorn, G. & Bynum, N. (1999). Tropical forest synergies. *Science (Washington DC)*, **286**, 2093–4.

Heaney, L. R. (1991). A synopsis of climate and vegetational change in Southeast Asia. *Climate Change*, **19**, 53–61.

Hedges, S. (1996). Proposal for inclusion of Banteng (*Bos javanicus*) in CITES. Appendix I. *IUCN/SSC Asian Wild Cattle Specialist Group, IUCN/SSC Wildlife Trade Programme*, Gland, Switzerland: Thai Government.

Heydon, M. J. & Bulloh, P. (1996). The impact of selective logging on sympatric civet species in Borneo. *Oryx*, **30**, 31–6.

Heydon, M. J. & Bulloh, P. (1997). Mousedeer densities in a tropical rainforest: the impact of selective logging. *Journal of Applied Ecology*, **34**, 484–96.

Heywood, V. H. & Stuart, S. N. (1992). Species extinctions in tropical forests. In *Tropical Deforestation and Species Extinction*, ed. T. C. Whitmore & J. A. Sayer. London: Chapman & Hall, pp. 91–117.

Heywood, V. H., Mace, G. M., May, R. M. & Stuart, S. N. (1994). Uncertainties in extinction rates. *Nature*, **368**, 105.

Hill, J. K. (1999). Butterfly spatial distribution and habitat requirements in a tropical forest: impacts of selective logging. *Journal of Applied Ecology*, **36**, 564–72.

Hill, J. K., Hamer, K. C., Lace, L. A. & Banham, W. M. T. (1995). Effects of selective logging on tropical forest butterflies on Buru, Indonesia. *Journal of Applied Ecology*, **32**, 754–60.

Hill, J. K., Hamer, K. C., Tangah, J. & Dawood, M. (2001). Ecology of tropical butterflies in rainforest gaps. *Oecologia*, **128**, 294–302.

Hill, J. K., Hamer, K. C., Dawood, M. M., Tangah, J. & Chey, V. K. (2003). Rainfall but not selective logging affect changes in abundance of a tropical forest butterfly in Sabah, Borneo. *Journal of Tropical Ecology*, **19**, 35–42.

Hilton, M. J. & Manning, S. S. (1995).
Conversion of coastal habitats in
Singapore: indications of unsustainable
development. *Environmental Conservation*,
22, 307–22.

Hoffman, W. A., Schroeder, W. & Jackson, R. B.
(2003). Regional feedbacks among fire,
climate, and tropical deforestation. *Journal
of Geophysical Research*, **108**, 4721.

Holloway, J. D. & Barlow, H. S. (1992). Potential
for loss of biodiversity in Malaysia, illustrated
by the moth fauna. In *Pest Management and
the Environment in 2000*, ed. H. S. Barlow.
Kuala Lumpur, Malaysia: CAB International
and Agricultural Institute of Malaysia,
pp. 293–311.

Holloway, J. D. & Hall, R. (1998). SE Asian
geology and biogeography: an introduction.
In *Biogeography and Geological Evolution of
SE Asia*, ed. R. Hall & J. D. Holloway.
Leiden: Backhuys, pp. 1–23.

Houghton, J. T. (2004). *Global Warming: the
Complete Coverage*. Cambridge: Cambridge
University Press.

Houlahan, J. E., Findlay, C. S., Schmidt, B. R.,
Meyers, A. H. & Kuzmin, S. L. (2000).
Quantitative evidence for global amphibian
population declines. *Nature*, **404**, 752–5.

Howlett, B. E. & Davidson, D. W. (2003). Effects
of seed availability, site conditions, and her-
bivory on pioneer recruitment after logging in
Sabah, Malaysia. *Forest Ecology and
Management*, **184**, 369–83.

Hughes, J. B., Daily, G. C. & Ehrlich, P. R.
(1997). Population diversity: its extent and
extinction. *Science*, **278**, 689–92.

Hughes, J. B., Round, P. D. & Woodruff, D. S.
(2003). The Indochinese–Sundaic faunal
transition at the Isthmus of Kra: an analysis
of resident forest bird species distributions.
Journal of Biogeography, **30**, 569–80.

Hunter, M. L., Jr (2001). *Fundamentals of
Conservation Biology*. Malden, MA:
Blackwell Science.

Huong, S. L. & Sodhi, N. S. (1997). Status of the
oriental magpie robin *Copsychus saularis* in
Singapore. *Malayan Nature Journal*, **50**,
347–54.

Huth, A. & Ditzer, T. (2001). Long-term impacts
of logging in a tropical rain forest: a
simulation study. *Forest Ecology and
Management*, **142**, 33–51.

Ickes, K. (2001). Hyper-abundance of native wild
pigs (*Sus scrofa*) in a lowland dipterocarp rain
forest of Peninsular Malaysia. *Biotropica*, **33**,
682–90.

Ickes, K. & Thomas, S. C. (2003). Native, wild
pigs (*Sus scrofa*) at Pasoh and their impacts
on the plant community. In *Pasoh: Ecology of
a Lowland Rain Forest of Southeast Asia*, ed.
T. Okuda, N. Manokaran, Y. Matsumoto,
K. Niyama, S. C. Thomas & P. S. Ashton.
Tokyo: Springer-Verlag, pp. 507–20.

Ickes, K., Dewalt, S. J. & Appanah, S. (2001).
Effects of native pigs (*Sus scrofa*) on woody
understorey vegetation in a Malaysian
lowland rain forest. *Journal of Tropical
Ecology*, **17**, 191–206.

Inger, R. F. (1966). The systematics and
zoogeography of the Amphibia of Borneo.
Fieldiana Zoology, **52**, 1–402.

Inigo-Elias, E. E. & Ramos, M. A. (1991). The
psittacine trade in Mexico. In *Neotropical
Wildlife Use and Conservation*, ed. J. G.
Robinson & K. H. Redford. Chicago:
University of Chicago Press, pp. 380–92.

Intachat, J., Holloway, J. D. & Speight, M. R.
(1997). The effects of different forest management practices on geometroid moth
populations and their diversity in Peninsular
Malaysia. *Journal of Tropical Forest Science*,
9, 411–30.

Intergovernmental Panel on Climate Change
[IPCC] (2001). *Third Assessment Report of the
Intergovernmental Panel on Climate Change*
(*Working Group I & II*). Cambridge:
Cambridge University Press.

International Tropical Timber Organization
[ITTO] (2003). *Annual Review and Assessment
of the World Timer Situation*. Yokohama,
Japan: International Tropical Timber
Organization.

International Union for the Conservation of
Nature and Natural Resources [IUCN]
(2000). *Red List of Threatened Species*. Gland,
Switzerland: International Union for the
Conservation of Nature and Natural
Resources.

International Union for the Conservation of
Nature and Natural Resources [IUCN]
(2003). *Red List of Threatened Species*
(www.redlist.org). Gland, Switzerland:
International Union for the Conservation of
Nature and Natural Resources.

International Union for the Conservation of
Nature and Natural Resources [IUCN]
(2004). *IUCN Red List of Threatened Species*
(www.redlist.org). Gland, Switzerland:
International Union for the Conservation
of Nature and Natural Resources.

Iremonger, S., Ravilious, C. & Quinton, T.
(1997). A statistical analysis of global forest

conservation. In *A Global Overview of Forest Conservation. Including: GIS files of Forests and Protected Areas, Version 2. CD-ROM*, ed. S. Iremonger, C. Ravilious & T. Quinton. Cambridge, UK: CIFOR and WCMC.

Jablonski, N. G. (1993). Quaternary environments and the evolution of primates in East Asia, with notes on two new specimens of fossil Cercopithecidae from China. *Folia Primatologica*, **60**, 118–32.

James, A. N., Gaston, K. J. & Balmford, A. (1999). Balancing the earth's accounts. *Nature*, **401**, 323–4.

James, A. N., Gaston, K. J. & Balmford, A. (2000). Why private institutions alone will not do enough to protect biodiversity. *Nature*, **404**, 120.

James, A. N., Gaston, K. J. & Balmford, A. (2001). Can we afford to conserve biodiversity? *BioScience*, **51**, 43–52.

Jang, C. J., Nishigami, Y. & Yanagisawa, Y. (1996). Assessment of global forest change between 1986 and 1993 using satellite-derived terrestrial net primary productivity. *Environmental Conservation*, **23**, 315–21.

Janzen, D. H. (2004). Setting up tropical biodiversity for conservation through non-damaging use: participation by parataxonomists. *Journal of Applied Ecology*, **41**, 181–7.

Jeeva, D., Bignell, D. E., Eggleton, P. & Maryati, M. (1999). Respiratory gas exchanges of termites from the Sabah (Borneo) assemblage. *Physiological Entomology*, **24**, 11–17.

Jenkins, M. (1992). Biological diversity. In *The Conservation Atlas of Tropical Forests: Africa*, ed. J. A. Sayer, C. S. Harcourt & N. M. Collins. New York: IUCN/Macmillan, pp. 26–32.

Jenkins, M. (2003). Prospects for biodiversity. *Science*, **302**, 1175–7.

Jepson, P. (2001). Global biodiversity plan needs to convince local policy-makers. *Nature*, **409**, 12.

Jepson, P., Jarvie, J. K., MacKinnon, K. & Monk, K. A. (2001). The end for Indonesia's lowland forests? *Science*, **292**, 859–61.

Johns, A. D. (1986). Effects of selective logging on the ecological organization of a Peninsular Malaysian rainforest avifauna. *Forktail*, **1**, 65–79.

Johns, A. D. (1987). The use of primary and selectively logged rainforest by Malaysian hornbills (Bucerotidae) and implications for their conservation. *Biological Conservation*, **40**, 179–90.

Johns, A. D. (1988). Malaysian hornbills and logging: some new observations. *Oriental Bird Club Bulletin*, **8**, 11–15.

Johns, A. D. (1989). Recovery of a Peninsular Malaysian rainforest avifauna following selective timber logging: the first twelve years. *Forktail*, **4**, 89–105.

Johns, A. G. (1996). Bird population persistence in Sabahan logging concessions. *Biological Conservation*, **75**, 3–10.

Johns, A. G. & Johns, B. G. (1995). Tropical forest primates and logging: long-term coexistence? *Oryx*, **29**, 205–11.

Johnston, R. F. & Janiga, M. (1995). *Feral Pigeons*. New York: Oxford University Press.

Jones, D. T., Susilo, F. X., Bignell, D. E. *et al.* (2003). Termite assemblage collapse along a land-use intensification gradient in lowland central Sumatra, Indonesia. *Journal of Applied Ecology*, **40**, 380–91.

Jones, M. J., Sullivan, M. S., Marsden, S. J. & Linsley, M. D. (2001). Correlates of extinction risk of birds from two Indonesian islands. *Biological Journal of the Linnean Society*, **73**, 65–79.

Jones, M. J., Marsden, S. J. & Linsley, M. D. (2003). Effects of habitat change and geographical variation on the bird communities of two Indonesian islands. *Biodiversity and Conservation*, **12**, 1013–32.

Kanchanasakha, B., Simcharoen, S. & Than, U. T. (1998). *Carnivores of mainland South East Asia*. Thailand: WWF.

Kang, N., Hails, C. J. & Sigurdsson, J. B. (1991). Nest construction and egg-laying in edible nest-swiftlets *Aerodramus* spp. and the implications for harvesting. *Ibis*, **133**, 170–7.

Katili, J. A. (1978). Past and present geotectonic position of Sulawesi, Indonesia. *Tectonophysics*, **45**, 289–322.

Kawanishi, K. & Sunquist, M. E. (2004). Conservation status of tigers in a primary rainforest of Peninsular Malaysia. *Biological Conservation*, **120**, 329–44.

Kelly, D. (1994). The evolutionary ecology of mast seeding. *Trends in Ecology & Evolution*, **9**, 465–70.

Kerr, J. T. & Burkey, T. V. (2002). Endemism, diversity, and the threat of tropical moist forest extinctions. *Biodiversity and Conservation*, **11**, 695–701.

Kershaw, A. P. (1978). Record of last interglacial–glacial cycle from northeastern Queensland. *Nature*, **272**, 159–61.

Kershaw, A. P., Penny, D., van der Kaars, S., Anshari, G. & Thamotherampillai, A. (2001).

Vegetation and climate in lowland Southeast Asia at the Last Glacial Maximum. In *Faunal and Floral Migrations and Evolution in SE Asia–Australasia*, ed. I. Davidson, I. Metcalfe, M. Morwood & J. M. B. Smith. Lisse: A. A. Balkema, pp. 227–36.

Kevan, P. G., Hussein, N. T., Hussey, N. & Wahid, M. B. (1986). The use of *Elaeidobius kamerunicus* for pollination of oil palm. *Planter*, **62**, 89–99.

Kidson, C., Indaratna, K. & Looareesuwan, S. (2000). The malaria cauldron of Southeast Asia: conflicting strategies of contiguous nation states. *Parassitologia*, **42**, 101–10.

Kinnaird, M. F. & O'Brien, T. G. (1998). Ecological effects of wildfire on lowland rainforest in Sumatra. *Conservation Biology*, **12**, 954–6.

Kinnaird, M. F. & O'Brien, T. G. (2001). Who's scratching whom? Reply to Whitten *et al. Conservation Biology*, **15**, 1459.

Kinnaird, M. F., O'Brien, T. & Bennett, L. (1998). Who's fiddling while Asia burns? *Wildlife Conservation*, **February** 9.

Kiss, A. (2004). Is community-based ecotourism a good use of biodiversity conservation funds. *Trends in Ecology & Evolution*, **19**, 232–7.

Klein, A.-M., Steffen-Dewenter, I. & Tscharntke, T. (2003a). Pollination of *Coffea canephora* in relation to local and regional agroforestry management. *Journal of Applied Ecology*, **40**, 837–45.

Klein, A.-M., Steffen-Dewenter, I. & Tscharntke, T. (2003b). Fruit set of highland coffee increases with the diversity of pollinating bees. *Proceedings of the Royal Society of London Series B Biological Sciences*, **270**, 955–61.

Knight, W. J. & Holloway, J. D. (eds.) (1990). *Insects and the Rain Forests of South East Asia (Wallacea)*. London: Royal Entomological Society.

Knop, E., Ward, P. I. & Wich, S. A. (2004). A comparison of orangutan density in a logged and unlogged forest on Sumatra. *Biological Conservation*, **20**, 183–8.

Koh, L. P. & Sodhi, N. S. (2004). Importance of reserves, fragments and parks for butterfly conservation in a tropical urban landscape. *Ecological Applications*, **14**, 1695–708.

Koh, L. P., Sodhi, N. S., Tan, H. T. W. & Peh, K. S.-H. (2002). Factors affecting the distribution of vascular plants, springtails, butterflies and birds on small tropical islands. *Journal of Biogeography*, **29**, 93–108.

Koh, L. P., Sodhi, N. S. & Brook, B. W. (2004a). Ecological correlates of extinction proneness in tropical butterflies. *Conservation Biology*, **18**, 1571–8.

Koh, L. P., Sodhi, N. S. & Brook, B. W. (2004b). Co-extinctions of tropical butterflies and their hostplants. *Biotropica*, **36**, 272–4.

Koh, L. P., Dunn, R. R., Sodhi, N. S. *et al.* (2004c). Species co-extinctions and the biodiversity crisis. *Science*, **305**, 1632–4.

Komdeur, J. (1996). Breeding of the Seychelles magpie robin *Copsychus sechellarum* and its implications for its conservation. *Ibis*, **138**, 485–98.

Kong, L., Yuen, B. & Sodhi, N. S. (1997). Nature and nurture, danger and delight: urban women's experiences of the natural world. *Landscape Research*, **22**, 245–66.

Kong, L., Yuen, B., Sodhi, N. S. & Briffett, C. (1999). The construction and experience of nature: perspectives of urban youths. *Tijdschrift Voor Economische en Sociale Geografie*, **90**, 3–16.

Kong, Y. C., Keung, W. M., Yip, T. T. *et al.* (1987). Evidence that epidermal growth factor is present in swiftlet's (*Collocalia*) nest. *Comparative Biochemistry and Physiology*, **87**, 221–6.

Kottelat, M. (2001). *Fishes of Laos*. Sri Lanka: WHT Publications (Pte) Ltd.

Krantz, G. W. & Poinar, G. O. (2004). Mite, nematodes and the multimillion dollar weevil. *Journal of Natural History*, **38**, 135–41.

Kremen, C., Niles, J. O., Dalton, M. G. *et al.* (2000). Economic incentives for rain forest conservation across scales. *Science*, **288**, 1828–32.

Kummer, D. M. & Turner, B. L. I. (1994). The human causes of deforestation in southeast Asia: the recurrent patterns is that of large-scale logging for exports, followed by agricultural expansion. *Bioscience*, **44**, 323–8.

Kuusipalo, J., Jafarsidik, Y., Adjers, G. & Tuomela, K. (1996). Population dynamics of tree seedlings in a mixed dipterocarp rainforest before and after logging and crown liberation. *Forest Ecology and Management*, **81**, 85–94.

Kuusipalo, J., Hadengganan, S., Adjers, G. & Sagala, A. P. S. (1997). Effect of gap liberation on the performance and growth of dipterocarp trees in a logged-over rainforest. *Forest Ecology and Management*, **92**, 209–19.

Lacki, M. J. & Lancia, R. A. (1983). Effects of wild pigs on beech growth in Great Smoky Mountains National Park USA. *Journal of Wildlife Management*, **50**, 655–9.

Laidlaw, R. K. (2000). Effects of habitat disturbance and protected areas on mammals of Peninsular Malaysia. *Conservation Biology*, **14**, 1639–48.

Lambert, F. (1991). The conservation of fig-eating birds in Malaysia. *Biological Conservation*, **58**: 31–40.

Lambert, F. R. (1992). The consequences of selective logging for Bornean lowland forest birds. *Philosophical Transactions of the Royal Society of London B Biological Sciences*, **335**, 443–57.

Lambert, F. R. & Collar, N. J. (2002). The future for Sundaic lowland forest birds: long-term effects of commercial logging and fragmentation. *Forktail*, **18**, 127–46.

Lammertink, M. (2004). A multiple-site comparison of woodpecker communities in Bornean lowland and hill forests. *Conservation Biology*, **18**, 746–57.

Laurance, W. F. (1999). Reflections on the tropical deforestation crisis. *Biological Conservation*, **91**, 109–17.

Laurance, W. F. (2000). Cut and run: the dramatic rise of transitional logging in the tropics. *Trends in Ecology & Evolution*, **15**, 433–4.

Laurance, W. F. (2004). The perils of payoff: corruption as a threat to global diversity. *Trends in Ecology & Evolution*, **19**, 399–401.

Laurance, W. F., Albernaz, A. K. M. & da Costa, C. (2001). Is deforestation accelerating in the Brazilian Amazon? *Environmental Conservation*, **28**, 305–11.

Leck, C. F. (1979). Avian extinctions in an isolated tropical wet-forest preserve, Ecuador. *Auk*, **96**, 343–52.

Lee, R. J. (2000). Market hunting pressures in north Sulawesi, Indonesia. *Tropical Biodiversity*, **6**, 145–62.

Lee, R. J., O'Brien, T. G., Kinnaird, M. F. & Dwiyahreni, A. A. (1999). *Impact of Wildlife hunting in Sulawesi, Indonesia, Technical Memorandum 4.* Bogor: PKA/Wildlife Conservation Society Indonesia Program.

Lee, T. M., Soh, M. C. K., Sodhi, N. S., Koh, L. P. & Lim, S. L.-H. (2005). Effects of habitat disturbance on mixed species bird flocks in a tropical sub-montane rainforest. *Biological Conservation*, **122**, 193–204.

Lekagul, B. & McNeely, J. A. (1988). *Mammals of Thailand*, 2nd edn. Bangkok: Saha Karn Bhaet.

Lessa, E. P., van Valkenburgh, B. & Farina, R. A. (1997). Testing hypotheses of differential mammalian extinctions subsequent to the Great American Biotic Interchange. *Palaeogeography, Palaeoclimatology, Palaeoecology*, **135**, 157–62.

Lessa, E. P., Cook, J. A. & Patton, J. L. (2003). Genetic footprints of demographic expansion in North America, but not Amazonia, during the Late Quaternary. *Proceedings of the National Academy of Sciences of the United States of America*, **100**, 10331–4.

Ling, S., Kumpel, N. & Albrechtsen, L. (2002). No new recipes for bushmeat. *Oryx*, **36**, 330.

Liow, L. H. (2000). Mangrove conservation in Singapore: a physical or psychological impossibility? *Biodiversity and Conservation*, **9**, 309–32.

Liow, L. H., Sodhi, N. S. & Elmqvist, T. (2001). Bee diversity along a disturbance gradient in tropical lowland forests of south-east Asia. *Journal of Applied Ecology*, **38**, 180–92.

Long, A., Collar, N., Wege, D. *et al.* (1996). Establishing conservation priorities using endemic birds. In *The Conservation Atlas of Tropical Forests: the Americas*, ed. C. S. Harcourt, J. A. Sayer & C. Billington. New York: IUCN/Simon & Schuster, pp. 35–46.

Long, A. J. (1994). The importance of tropical montane cloud forests for endemic and threatened birds. In *Tropical Montane Cloud Forests*, ed. L. S. Hamilton, J. S. Juvik & F. N. Scatena. New York: Springer-Verlag, pp. 79–106.

Lovejoy, T. E., Bierregaard, R. O. J., Rylands, A. B., Malcolm, J. R. & Quintela, C. E. (1997). Edge and other effects of isolation on Amazon forest fragments. In *Tropical Forest Remnants: Ecology, Management and Conservation of Fragmented Communities*, ed. W. F. Laurance & R. O. J. Bierregaard. Chicago: University of Chicago Press, pp. 257–85.

MacDonald, J. A., Jeeva, D., Eggleton, P. *et al.* (1999). The effect of termite biomass and anthropogenic disturbance on the CH_4 budgets of tropical forests in Cameroon and Borneo. *Global Change Biology*, **5**, 869–79.

Mace, G. M., Balmford, A., Boitani, L. *et al.* (2000). It's time to work together and stop duplicating conservation efforts. *Nature*, **405**, 393.

MacKinnon, J. & Phillipps, K. (1993). *The Birds of Borneo, Sumatra, Java and Bali.* Oxford: Oxford University Press.

Madge, S. & McGowan, P. (2002). *Pheasants, Partridges & Grouse.* London: Christopher Helm.

Mainka, S. A. & Mills, J. A. (1995). Wildlife and traditional Chinese medicine – supply and demand for wildlife species. *Journal of Zoo and Wildlife Medicine*, **26**, 193–200.

Maley, J. (1987). Fragmentation de la forêt dense humid africaine et extension des biotypes montagnards au Quaternaire récent: nouvelles données polliniques et chronologiques. Implications paléoclimatiques et biogéographiques. *Palaeoecology of Africa*, **18**, 307–44.

Malhi, Y. & Grace, J. (2000). Tropical forests and atmospheric carbon dioxide. *Trends in Ecology & Evolution*, **15**, 332–7.

Mann, C. (1990). Meta-analysis in the breech. *Science*, **249**, 476–80.

Marod, D., Kutintara, U., Tanaka, H. & Nakashizuka, T. (2002). The effects of drought and fire on seed and seedling dynamics in a tropical seasonal forest in Thailand. *Plant Ecology*, **161**, 41–57.

Marsden, S. J. (1998). Changes in bird abundance following selective logging on Seram, Indonesia. *Conservation Biology*, **12**, 605–11.

Marsden, S. J. & Jones, M. J. (1997). The nesting requirements of the parrots and hornbill of Sumba, Indonesia. *Biological Conservation*, **82**, 279–87.

Matthews, E. (2001). *Understanding the FRA 2000. World Resources Institute Forest Briefing No. 1*. Washington: World Resources Institute.

Mayr, E. (1941). The origin and the history of the bird fauna of Polynesia. *Proceedings of the 6th Pacific Science Congress*, **4**, 197–216.

McConkey, K. R. (2000). Primary seed shadow generated by gibbons in the rain forests of Barito Ulu, central Borneo. *American Journal of Primatology*, **52**, 13–29.

McGowan, P. J. K. & Garson, P. J. (2002). The Galliformes are highly threatened: should we care? *Oryx*, **36**, 311–12.

McKinney, M. L. (1997). Extinction vulnerability and selectivity: combining ecological and paleontological views. *Annual Review of Ecology and Systematics*, **28**, 495–516.

McLain, D. K., Moulton, M. P. & Sanderson, J. G. (1999). Sexual selection and extinction: the fate of plumage-dimorphic and plumage-monomorphic birds introduced onto islands. *Evolutionary Ecology Research*, **1**, 549–65.

McNeely, J., Heywood, V., Jackson, P. W. *et al.* (1991). Forest wildlife. In *The Conservation Atlas of Tropical Forests – Asia and The Pacific*, ed. N. M. Collins, J. A. Sayer & T. C. Whitmore. London: IUCN/Macmillan Press Ltd, pp. 13–24.

Medway, G. & Wells, D. R. (1976). *The Birds of the Malay Peninsula*. London: H. F. & G. Witherby Ltd.

Meijaard, E. (2003). Mammals of south-east Asian islands and their Late Pleistocene environments. *Journal of Biogeography*, **30**, 1245–57.

Meijaard, E. (2004). Biogeographic history of the Javan leopard *Panthera pardus* based on a craniometric analysis. *Journal of Mammalogy*, **85**, 302–10.

Meijaard, E. & Nijman, V. (2000). The local extinction of the proboscis monkey *Nasalis larvatus* in Pulau Kaget Nature Reserve, Indonesia. *Oryx*, **34**, 66–70.

Mendelsohn, R. & Balick, M. (1995). Private property and rainforest conservation. *Conservation Biology*, **9**, 1322–3.

Metcalfe, I. (1998). Palaeozoic and Mesozoic geological evolution of the SE Asian region: multidisciplinary constraints and implications for biogeography. In *Biogeography and Geological Evolution of SE Asia*, ed. R. Hall & J. D. Holloway. Leiden: Backhuys, pp. 25–41.

Michaux, B. (1998). Terrestrial birds of the Indo-Pacific. In *Biogeography and Geological Evolution of SE Asia*, ed. R. Hall & J. D. Holloway. Leiden: Backhuys, pp. 361–91.

Mills, J. A. (1993). Tiger bone trade in South Korea. *Cat News*, **19**, 13–16.

Milner-Gulland, E. J. & Bennett, E. L. (2003). Wild meat: the bigger picture. *Trends in Ecology & Evolution*, **18**, 351–7.

Mirmanto, E., Proctor, J., Green, J., Nagy, L. & Suriantata (1999). Effects of nitrogen and phosphorus fertilization in a lowland evergreen rainforest. *Philosophical Transactions of the Royal Society of London B Biological Sciences*, **354**, 1825–9

Mitra, S. S. & Sheldon, F. H. (1993). Use of an exotic tree plantation by Bornean lowland forest birds. *Auk*, **110**, 529–40.

Mittermeier, R. A., Myers, N., Thomsen, J. B., da Fonseca, G. A. B. & Olivieri, S. (1998). Biodiversity hotspots and major tropical wilderness areas: approaches to setting conservation priorities. *Conservation Biology*, **12**, 516–20.

Mittermeier, R. A., Myers, N., Gil, P. R. & Mittermeier, C. G. (1999). *Hotspots: Earth's Biologically Richest and Most Endangered Terrestrial Ecoregions*. Monterrey, Mexico: Cemex, Conservation International and Agrupacion Sierra Madre.

Moran, E. F. (1988). Following the Amazonian highways. In *People of the Rain Forest*, ed.

J. S. Denslow & C. Padoch. Berkeley: University of California Press pp. 155–66.

Morley, R. J. (1998). Palynological evidence for tertiary plant dispersals in the SE Asia region in relation to plate tectonics and climate. In *Biogeography and Geological Evolution of SE Asia*, ed. R. Hall & J. D. Holloway. Leiden: Backhuys, pp. 177–200.

Morley, R. J. (2000). *Origin and Evolution of Tropical Rain Forests*. Chichester, England: John Wiley & Sons Ltd.

Morrogh-Bernard, H., Husson, S., Page, S. E. & Rieley, J. O. (2003). Population status of the Bornean orang-utan (*Pongo pygmaeus*) in the Sebangau peat swamp forest, central Kalimantan, Indonesia. *Biological Conservation*, **110**, 141–52.

Musser, G. G. & Newcomb, C. (1983). Malaysian murids and the giant rat of Sumatra. *Bulletin of the American Museum of Natural History*, **174**, 329–598.

Myers, N. (1986). Tropical deforestation and a mega-extinction spasm. In *Conservation Biology: The Science of Scarcity and Diversity*, ed. M. E. Soulé. Sunderland, MA: Sinauer, pp. 394–409.

Myers, N. (1991). Tropical forests present status and future outlook. *Climatic Change*, **19**, 3–32.

Myers, N. (1996). Environmental services of biodiversity. *Proceedings of the National Academy of Sciences of the United States of America*, **93**, 2764–9.

Myers, N., Mittermeier, R. A., Mittermeier, C. G., da Fonseca, G. A. B. & Kent, J. (2000). Biodiversity hotspots for conservation priorities. *Nature*, **403**, 853–8.

Newsome, J. & Flenley, J. R. (1988). Late Quaternary vegetational history of the central highlands of Sumatra. II. Palaeopalynology and vegetational history. *Journal of Biogeography*, **15**, 555–78.

Ng, P. K. L., Chou, L. M. & Lam, T. J. (1993). The status and impact of introduced freshwater animals in Singapore. *Biological Conservation*, **64**, 19–24.

Niklaus, P. A., Leadley, P. W., Schmid, B. & Korner, C. (2001). A long-term field study on biodiversity x elevated CO_2 interactions in grassland. *Ecological Monographs*, **71**, 341–56.

Norris, D. R. (2004). Mosquito-borne diseases as a consequence of land use change. *EcoHealth*, **1**, 19–24.

Norris, K. & Harper, N. (2004). Extinction processes in hot spots of avian biodiversity and the targeting of pre-emptive conservation action. *Proceedings of the Royal Society of London Series B Biological Sciences*, **271**, 123–30.

Novacek, M. J. (2001). *The Biodiversity Crisis: Losing What Counts*. New York: The New Press.

Nowell, K., Chyi, W.-L. & Pei, C.-J. (1992). *The Horns of a Dilemma: The Market for Rhino Horn in Taiwan*. Cambridge, UK: TRAFFIC International.

Nyhus, P. J., Tilson, R. & Sumianto (2000). Crop-raiding elephants and conservation implications at Way Kambas National Park, Sumatra, Indonesia. *Oryx*, **34**, 262–74.

O'Brien, T. G. & Kinnaird, M. F. (1996). Changing populations of birds and mammals in north Sulawesi. *Oryx*, **30**, 150–6.

O'Brien, T. G. & Kinnaird, M. F. (2003). Caffeine and conservation. *Science*, **300**, 587.

O'Brien, T. G., Kinnaird, M. F., Nurcahyo, A., Prasetyaningrum, M. & Iqbal, M. (2003). Fire, demography and the persistence of siamang (*Symphalangus syndactylus*: Hylobatidae) in a Sumatran rainforest. *Animal Conservation*, **6**, 115–21.

Ohsawa, M. (1995). The montane cloud forest and its gradational changes in Southeast Asia. In *Tropical Montane Cloud Forests*, ed. L. S. Hamilton, J. S. Juvik & F. N. Scatena. New York: Springer-Verlag, pp. 254–65.

Oka, T., Iskandar, E. & Ghozali, D. I. (2000). Effects of forest fragmentation on the behavior of Bornean gibbons. In *Rainforest Ecosystems of East Kalimantan: El Nino, Drought, Fire and Human Impacts. Ecological Studies Vol. 140*, ed. E. Guhardja, M. Fatawi, M. Sutisna, T. More & S. Ohta. Tokyo: Springer-Verlag, pp. 229–41.

Okuda, T., Suzuki, M., Adachi, N. *et al.* (2003). Effect of selective logging on canopy and stand structure and tree species composition in a lowland dipterocarp forest in Peninsular Malaysia. *Forest Ecology and Management*, **175**, 297–320.

Ong, L. (2000). Determining conservation needs of four key vertebrate in Southeast Asia. MSc thesis. National University of Singapore.

Ooi, J. B. (1976). *Peninsular Malaysia*. London: Longman.

Ooi, J. B. (1990). The tropical rain forest: patterns of exploitation and trade. *Singapore Journal of Tropical Geography*, **11**, 117–42.

Page, S. E. & Rieley, J. O. (1998). Tropical peatlands: a review of their natural resource functions with particular reference to

Southeast Asia. *International Peat Journal*, **8**, 95–106.

Page, S. E., Siegert, F., Rieley, J. O. *et al.* (2002). The amount of carbon released from peat and forest fires in Indonesia during 1997. *Nature*, **420**, 61–5.

Palmer, M., Bernhardt, E., Chornesky, E. *et al.* (2004). Ecology for a crowded planet. *Science*, **304**, 1251–2.

Parmesan, C., Ryrholm, N., Stefanescu, C. *et al.* (1999). Poleward shifts in geographical ranges of butterfly species associated with regional warming. *Nature*, **399**, 579–83.

Pattanavibool, A. & Dearden, P. (2002). Fragmentation and wildlife in montane ever-green forests, northern Thailand. *Biological Conservation*, **107**, 155–64.

Pattanavibool, A. & Edge, W. D. (1996). Single-tree selection silviculture affects cavity resources in mixed deciduous forests in Thailand. *Journal of Wildlife Management*, **60**, 67–73.

Payne, J., Francis, C. M. & Phillipps, K. (1985). *A Field Guide to the Mammals of Borneo*. Kuala Lumpur, Malaysia: The Sabah Society, Kota Kinabalu, Sabah, and WWF Malaysia.

Pearson, H. (2003). Lost forest fuels malaria. *Nature Science Update* (www.nature.com/nsu/031124/031124–12.html).

Peh, K. S.-H., de Jong, J., Sodhi, N. S., Lim, S. & Yap, C. A.-M. (2005). Lowland rainforest avifauna and human disturbance: persistence of primary forest birds in selectively logged forest and mixed-rural habitats of southern Peninsular Malaysia. *Biological Conservation*, **123**, 489–505.

Pell, A. S. & Tidemann, C. R. (1997). The ecology of the common myna in urban nature reserves in the Australian Capital Territory. *Emu*, **97**, 141–9.

Peters, H. A. (2001). *Clidemia hirta* invasion at the Pasoh Forest Reserve: an unexpected plant invasion in an undisturbed tropical forest. *Biotropica*, **33**, 60–8.

Peters, R. H. (1983). *The Ecological Implications of Body Size*. New York: Cambridge University Press.

Phat, N. K., Knorr, W. & Kim, S. (2004). Appropriate measures for conservation of terrestrial carbon stocks – analysis of trends of forest management in southeast Asia. *Forest Ecology and Management*, **191**, 283–99.

Pimental, D., Lach, L., Zuniga, R. & Morrison, D. (2000). Environmental and economic costs of nonindigenous species in the United States. *Bioscience*, **50**, 53–65.

Pimm, S. L. & Kitching, R. L. (1988). Food web patterns: trivial flaws or the basis of an active research program? *Ecology*, **69**, 1669–72.

Pimm, S. L., Lawton, J. H. & Cohen, J. E. (1991). Food web patterns and their consequences. *Nature*, **350**, 669–74.

Pimm, S. L. & Raven, P. (2000). Extinction by numbers. *Nature*, **403**, 843–5.

Pimm, S., Russell, G., Gittleman, J. & Brooks, T. (1995). The future of biodiversity. *Science*, **269**, 347–54.

Pinard, M., Howlett, B. & Davidson, D. (1996). Site conditions limit pioneer tree recruitment after logging of dipterocarp forests in Sabah, Malaysia. *Biotropica*, **28**, 2–12.

Platt, S. G. & Tri, N. V. (2000). Status of the Siamese crocodile in Vietnam. *Oryx*, **34**, 217–21.

Poonswad, P., Sukkasem, C., Phataramata, S. *et al.* (2005). Comparison of cavity modifica-tion and community involvement as strategies for hornbill conservation in Thailand. *Biological Conservation*, **122**, 385–93.

Posey (ed.) (1999). *Cultural and Spiritual Value of Biodiversity*. London: United Nations Environment Programme and Intermediate Technology Publication.

Pratt, D. G., Macmillan, D. C. & Gordon, I. J. (2004). Local community attitudes to wildlife utilisation in the changing economic and social context of Mongolia. *Biodiversity and Conservation*, **13**, 591–613.

Prawiradilaga, D. M. (1997). The maleo *Macrocephalon maleo* on Buton. *Bulletin of the British Ornithologists' Club*, **117**, 237.

Primack, R. B. (2000). *A Primer of Conservation Biology*. Sunderland, MA: Sinauer Associates Inc.

Purvis, A., Agapow, P. M., Gittleman, J. L. & Mace, G. M. (2000). Nonrandom extinction and the loss of evolutionary history. *Science*, **288**, 328–30.

Putz, F. E., Dykstra, D. P. & Heinrich, R. (2000). Why poor logging practices persist in the tropics. *Conservation Biology*, **14**, 951–6.

Rabinowitz, A. (1995). Helping a species go extinct – the Sumatran rhino in Borneo. *Conservation Biology*, **9**, 482–8.

Rambo, A. (1979). Primitive man's impact on genetic resources of the Malaysian tropical rain forest. *Malaysian Applied Biology*, **8**, 59–65.

Randrianasolo, A., Miller, J. S. & Consiglio, T. K. (2002). Application of IUCN criteria

and Red List categories to species of
five Anacardiaceae genera in
Madagascar. *Biodiversity and Conservation*,
11, 1289–300.

Rangin, C., Pubellier, M., Azema, J. *et al.* (1990).
The quest for Tethys in the western Pacific –
8 paleogeodynamic maps for Cenozoic time.
Bulletin de la Société Géologique de France, **8**,
909–13.

Rao, M. & van Schaik, C. P. (1997). The beha-
vioral ecology of Sumatran orangutans in
logged and unlogged forest. *Tropical
Biodiversity*, **4**, 173–85.

Rao, M., Rabinowitz, A. & Khaing, S. T. (2002).
Status review of the protected-area system
in Myanmar, with recommendations for
conservation planning. *Conservation Biology*,
16, 360–8.

Reaka-Kudla, M. L., Wilson, D. E. & Wilson,
E. O. (1997). *Biodiversity II: Understanding
and Protecting our Biological Resources.*
Washington, DC: Joseph Henry Press.

Redford, K. H., Naughton, L., Raez-Luna, E.,
Gimenez-Dixon, M. & Sizer, N. (1996).
Forest wildlife and its exploitation by
humans. In *The Conservation Atlas of
Tropical Forests: the Americas*, ed.
C. S. Harcourt, J. A. Sayer & C. Billington. New
York: IUCN/Simon & Schuster, pp. 47–56.

Reed, J. M. (1999). The role of behaviour in
recent avian extinctions and endangerments.
Conservation Biology, **13**, 232–41.

Richards, J. F. & Flint, E. P. (1994). A century of
land-use change in South and Southeast Asia.
In *Effects of Land-use Change on Atmospheric
CO₂ Concentrations*, ed. V. H. Dale. New
York: Springer-Verlag, pp. 15–66.

Richards, P. W. (1973). Africa, the 'odd man
out'. In *Tropical Forest Ecosystems of
Africa and South America: a Comparative
Review*, ed. B. J. Meggers, E. S. Ayensu &
W. D. Duckworth. Washington, DC:
Smithsonian Institution Press, pp. 21–6.

Ricketts, T. H., Daily, G. C., Ehrlich, P. R. &
Michener, C. D. (2004). Economic value
of tropical forest to coffee production.
*Proceedings of the National Academy of
Sciences of the United States of America*, **34**,
12579–82.

Ridder-Numan, J. (1998). Historical
biogeography of *Spatholobus*
(Leguminonsae-Papilionoideae) and
allies in Asia. In *Biogeography and
Geological Evolution of SE Asia*, ed. R. Hall &
J. D. Holloway. Leiden: Backhuys,
pp. 259–77.

Riley, J. (2002). Mammals on the Sangihe and
Talaud Islands, Indonesia, and the impact
of hunting and habitat loss. *Oryx*, **36**,
288–96.

Robertson, J. M. Y. & van Schaik, C. P. (2001).
Causal factors underlying the dramatic
decline of the Sumatran orang-utan. *Oryx*,
35, 26–38.

Robinson, J. G. & Bennett, E. L. (2002). Will
alleviating poverty solve the bushmeat crisis?
Oryx, **36**, 332.

Robinson, J. G., Redford, K. H. & Bennett, E. L.
(1999). Wildlife harvest in logged tropical
forests. *Science*, **284**, 595–6.

Robinson, J. M. (2001). The dynamics of avicul-
tural markets. *Environmental Conservation*,
28, 76–85.

Rodrigues, A. S. L., Andelman, S. J., Bakarr,
M. I. *et al.* (2004). Effectiveness of the global
protected area network in representing
species diversity. *Nature*, **428**, 640–3.

Rosenbaum, B., O'Brien, T. G., Kinnaird, M. &
Supriatna, J. (1998). Population densities of
Sulawesi crested black macaques (*Macaca
nigra*) on Bacan and Sulawesi, Indonesia:
effects of habitat disturbance and hunting.
American Journal of Primatology, **44**, 89–106.

Round, P. D. & Brockelman, W. Y. (1998). Bird
communities in disturbed lowland forest
habitats of southern Thailand. *Natural
History Bulletin of the Siam Society*, **46**,
171–96.

Rowcliffe, M. (2002). Bushmeat and the biology
of conservation. *Oryx*, **36**, 331.

Ruangpanit, N. (1995). Tropical seasonal
forests in monsoon Asia: with emphasis
on continental Southeast Asia. *Vegetatio*,
121, 31–40.

Ruedi, M. & Fumagalli, L. (1996). Genetic
structure of gymnures (genus *Hylomys*;
Erinaceidae) on continental islands of
Southeast Asia: historical effects of fragment-
ation. *Journal of Zoological Systematics and
Evolutionary Research*, **34**, 153–62.

Ryall, C. (1994). Recent extensions of range in the
house crow *Corvus splendens. Bulletin of the
British Ornithologists' Club*, **114**, 90–100.

Sakagami, S. F., Inoue, T. & Salmah, S. (1990).
Stingless bees of Sumatra. In *Natural History
of Social Wasps and Bees in Equatorial
Sumatra*, ed. S. F. Sakagami, R. Ohgushi &
D. W. Roubik. Sapporo, Japan: Hokkaido
University Press, pp. 125–38.

Sala, O. E., Chapin, F. S., 3rd & Armesto, J. J.
et al. (2000). Global biodiversity scenarios for
the year 2100. *Science*, **287**, 1770–4.

Samejima, H., Marzuki, M., Nagamitsu, T. & Nakasizuka, T. (2004). The effects of human disturbance on a stingless bee community in a tropical rainforest. *Biological Conservation*, 120, 577–87.

Sasekumar, A., Ong, J. E. & Wilkinson, C. R. (1994). Action agenda for ASEAN mangroves: now and into the future. In *Proceedings of the 3rd ASEAN–Australia Symposium on Living Coastal Resources, 16–20 May 1994, Bangkok*, ed. C. R. Wilkinson, S. Sudara & L. M. Chou. Townsville: Australian Institute of Marine Science, pp. 98–103.

Sastry, N. (2002). Forest fires, air pollution, and mortality in Southeast Asia. *Demography*, 39, 1.

Saw, S.-H. (1970). *Singapore: Population in Transition*. Philadelphia: University of Pennsylvania Press.

Schafer, E. H. (1963). *The Golden Peaches of Samarkand: A Study of T'ang Exotics*. Berkeley: University of California Press.

Schweithelm, J. (1998). *The Fire This Time. An Overview of Indonesia's Forest Fires in 1997/1998*. Jakarta, Indonesia: WWF.

Seidensticker, J., Christie, S. & Jackson, P. (1999). *Riding the Tiger: Tiger Conservation In Human-dominated Landscapes*. Cambridge, UK: Cambridge University Press.

Sekercioglu, C. H., Ehrlich, P. R., Daily, G. C. *et al.* (2002). The disappearance of insectivorous birds from tropical forest fragments. *Proceedings of the National Academy of Sciences of the United States of America*, 99, 263–7.

Servant, M., Maley, J., Turcq, B. *et al.* (1993). Tropical forest changes during the Late Quaternary in African and South American lowlands. *Global Planetary Change*, 7, 25–40.

Shahabuddin, G., Herzner, G. A., Aponte, C. R. & Gomez, M. D. C. (2000). Persistence of a frugivorous butterfly species in Venezuelen forest fragments: the role of movement and habitat quality. *Biodiversity and Conservation*, 9, 1623–41.

Shapiro, A. M. (2002). The Californian urban butterfly fauna is dependent on alien plants. *Diversity and Distributions*, 8, 31–40.

Sheil, D. & Lawrence, A. (2004). Tropical biologists, local people and conservation: new opportunities for collaboration. *Trends in Ecology & Evolution*, 19, 634–8.

Sheil, D., Sayer, J. A. & O'Brien, T. (1999). Tree species diversity in logged rainforests. *Science*, 284, 1587.

Shepherd, C. R. & Magnus, N. (2004). *Nowhere to Hide: The Trade in Sumatran Tiger. A TRAFFIC Southeast Asia Report*. Selangor, Malaysia: WWF, Fauna and Flora International, Wildlife Conservation Society.

Shi, G. R. & Archibald, N. W. (1998). Permian marine biogeography of SE Asia. In *Biogeography and Geological Evolution of SE Asia*, ed. R. Hall & J. D. Holloway. Leiden: Backhuys, pp. 57–72.

Shine, R., Ambariyanto, Harlow, P. S. & Mumpuni (1999). Reticulated pythons in Sumatra: biology, harvesting and sustainability. *Biological Conservation*, 87, 349–57.

Sim, J. & Soares, E. (2003). *Killer carriers? Streats* (27 May 2003).

Simberloff, D. (1992). Do species–area curves predict extinction in fragmented forests? In *Tropical Deforestation and Species Extinction*, ed. T. C. Whitmore & J. A. Sayer. London: Chapman & Hall, pp. 75–89.

Simberloff, D. & Cox, J. (1987). Consequences and costs of conservation corridors. *Conservation Biology*, 1, 63–71.

Singleton, I. & van Schaik, C. P. (2001). Orangutan home range size and its determinants in a Sumatran swamp forest. *International Journal of Primatology*, 22, 877–911.

Sist, P., Sheil, D., Kartawinata, K. & Priyadi, H. (2003). Reduced-impact logging in Indonesian Borneo: some results confirming the need for new silvicultural prescriptions. *Forest Ecology and Management*, 179, 415–27.

Sizer, N. & Plouvier, D. (2000). *Increased Investment and Trade by Transnational Logging Companies in Africa, the Caribbean and the Pacific: Implications for the Sustainable Management and Conservation of Tropical Forests*. WWF, WRI and EU.

Slik, J. W. F., Verburg, R. W. & Kessler, P. J. A. (2002). Effects of fire and selective logging on the tree species composition of lowland dipterocarp forest in East Kalimantan, Indonesia. *Biodiversity and Conservation*, 11, 85–98.

Smith, R. J., Muir, R. D. J., Walpole, M. J., Balmford, A. & Leader-Williams, N. (2003). Governance and the loss of biodiversity. *Nature*, 426, 67–70.

Smith, S. D., Huxman, T. E., Zitzer, S. F. *et al.* (2000). Elevated CO_2 increases productivity and invasive species success in an arid ecosystem. *Nature*, 408, 79–82.

Smythies, B. E. (1953). *The Birds of Burma.* London: Oliver and Boyd.

Sodhi, N. S. (2002). A comparison of bird communities of two fragmented and two continuous Southeast Asian rainforests. *Biodiversity and Conservation*, **11**, 1105–19.

Sodhi, N. S. & Er, K. B. H. (2000). Conservation meets consumption. *Trends in Ecology & Evolution*, **15**, 431.

Sodhi, N. S. & Liow, L. H. (2000). Improving conservation biology research in Southeast Asia. *Conservation Biology*, **14**, 1211–12.

Sodhi, N. S. & Sharp, I. (2005). *Winged Invaders: Pest Birds of Asia-Pacific.* Singapore: SNP Press.

Sodhi, N. S., Peh, K. S. -H., Lee, T. M. *et al.* (2003). Artificial nest and seed predation experiments on tropical southeast Asian islands. *Biodiversity and Conservation*, **12**, 2415–33.

Sodhi, N. S., Briffett, C., Kong, L. & Yuen, B. (1999). Bird use of linear areas of a tropical city: implications for park connector design and management. *Landscape and Urban Planning*, **45**, 123–30.

Sodhi, N. S., Liow, L. H. & Bazzaz, F. A. (2004a). Avian extinctions from tropical and subtropical forests. *Annual Review of Ecology, Evolution and Systematics*, **35**, 323–45.

Sodhi, N. S., Koh, L. P., Brook, B. W. & Ng, P. K. L. (2004b). Southeast Asian biodiversity: an impending disaster. *Trends in Ecology & Evolution*, **19**, 654–60.

Sodhi, N. S., Lee, T. M., Koh, L. P. & Dunn, R. R. (2005a). A century of avifaunal losses from a small tropical rainforest fragment. *Animal Conservation*, **8**, 217–22.

Sodhi, N. S., Soh, M. C. K., Prawiradilaga, D. M. & Brook, B. W. (2005b). Persistence of lowland forest birds in a recently logged area in central Java. *Bird Conservation International*, **15**, 173–91.

Sodhi, N. S., Koh, L. P., Prawiradilaga, D. M. *et al.* (2005c). Land use and conservation value for forest birds in central Sulawesi (Indonesia). *Biological Conservation*, **122**, 547–58.

Spencer, J. E. (1966). *Shifting Cultivation in Southeastern Asia.* Berkeley, California: University of Chicago Press.

Spitzer, K., Novotny, V., Tonner, M. & Leps, J. (1993). Habitat preferences, distribution and seasonality of the butterflies (Lepidoptera, Papilionoidea) in a montane tropical rainforest, Vietnam. *Journal of Biogeography*, **20**, 109–21.

Steppan, S. J., Zawadzki, C. & Heaney, L. R. (2003). Molecular phylogeny of the endemic Philippine rodent *Apomys* (Muridae) and the dynamics of diversification in an oceanic archipelago. *Biological Journal of the Linnean Society*, **80**, 699–715.

Stibig, H.-J. & Malingreau, J.-P. (2003). Forest cover of insular Southeast Asia mapped from recent satellite images of coarse spatial resolution. *Ambio*, **32**, 469–75.

Stolle, F., Chomitz, K. M., Lambin, E. F. & Tomich, T. P. (2003). Land use and vegetation fires in Jambi Province, Sumatra, Indonesia. *Forest Ecology and Management*, **179**, 277–92.

Stork, N. E. & Lyal, C. H. C. (1993). Extinction or 'co-extinction' rates? *Nature*, **366**, 307.

Stuijts, I. M. (1993). Late Pleistocene and Holocene vegetation of West Java, Indonesia. *Modern Quaternary Research in Southeast Asia*, **12**, 1–173.

Styring, A. R. & Hussin, M. Z. B. (2004). Effects of logging on woodpeckers in a Malaysian rain forest: the relationship between resource availability and woodpecker abundance. *Journal of Tropical Ecology*, **20**, 495–504.

Styring, A. R. & Ickes, K. (2001). Woodpecker abundance in a logged (40 years ago) vs. unlogged lowland dipterocarp forest in Peninsular Malaysia. *Journal of Tropical Ecology*, **17**, 261–8.

Styring, A. R. & Ickes, K. (2003). Woodpeckers (Picidae) at Pasoh: foraging ecology, flocking and the impacts of logging on abundance and diversity. In *Pasoh: Ecology of a Lowland Rain Forest of Southeast Asia*, ed. T. Okuda, N. Manokaran, Y. Matsumoto, K. Niyama, S. C. Thomas & P. S. Ashton. Tokyo: Springer-Verlag, pp. 547–57.

Sun, X., Li, X., Luo, Y. & Chen, X. (2000). The vegetation and climate at the last glaciation on the emerged continental shelf of the South China Sea. *Palaeogeography, Palaeoclimatology, Palaeoecology*, **160**, 301–16.

Syed, R. A., Law, I. H. & Corley, R. H. V. (1982). Insect pollination of oil palm: introduction, establishment and pollinating efficiency of *Elaeidobius kamerunicus* in Malaysia. *Planter*, **58**, 547–61.

Takamura, K. (2003). Is the termite community disturbed by logging? In *Pasoh: Ecology of a Lowland Rain Forest of Southeast Asia*, ed. T. Okuda, N. Manokaran, Y. Matsumoto, K. Niyama, S. C. Thomas & P. S. Ashton. Tokyo: Springer-Verlag, 521–31.

Talbott, K. & Brown, M. (1998). Forest plunder in Southeast Asia: an environmental security nexus in Burma and Cambodia. *Environmental Change and Security Project Report*, **4**, 53–60.

Taylor, D., Saksena, P., Sanderson, P. G. & Kucera, K. (1999). Environmental change and rain forests on the Sunda shelf of Southeast Asia: drought, fire and the biological cooling of biodiversity hotspots. *Biodiversity and Conservation*, **8**, 1159–77.

Teo, D. H. L., Tan, H. T. W., Corlett, R. T., Wong, C. M. & Lum, S. K. Y. (2003). Continental rain forest fragments in Singapore resist invasion by exotic plants. *Journal of Biogeography*, **30**, 305–10.

Terborgh, J. (1974). Preservation of natural diversity: the problem of extinction prone species. *Bioscience*, **24**, 715–22.

Terborgh, J. (1992). Maintenance of diversity in tropical forests. *Biotropica*, **24**, 283–92.

Terborgh, J., Lopez, L. & Jose-Tello, S. (1997). Bird communities in transition: the Lago Guri islands. *Ecology*, **78**, 1494–501.

Thiollay, J. M. (1989). Area requirements for the conservation of rain forest raptors and game birds in French Guiana. *Conservation Biology*, **3**, 128–37.

Thiollay, J.-M. (1995). The role of traditional agroforests in the conservation of rain forest bird diversity in Sumatra. *Conservation Biology*, **9**, 335–53.

Thiollay, J.-M. & Meyburg, B. U. (1988). Forest fragmentation and the conservation of raptors: a survey on the island of Java, Indonesia. *Biological Conservation*, **44**, 229–50.

Thomas, J. A. (1991). Rare species conservation: case of European butterflies. In *The Scientific Management of Temperate Communities for Conservation*, ed. I. Spellerberg & B. Goldsmith. Oxford, UK: Blackwell, pp. 149–97.

Thomas, J. A. & Morris, M. G. (1994). Patterns, mechanisms and rates of decline among UK invertebrates. *Philosophical Transactions of the Royal Society of London B Biological Sciences*, **344**, 47–54.

Thomas, J. A., Bourn, N. A. D., Clarke, R. T. *et al.* (2001). The quality and isolation of habitat patches both determine where butterflies persist in fragmented landscapes. *Proceedings of the Royal Society of London Series B Biological Sciences*, **268**, 1791–6.

Thomas, J. A., Telfer, M. G., Roy, D. B. *et al.* (2004). Comparative losses of British butterflies, birds, and plants and the global extinction crisis. *Science*, **303**, 1879–81.

Thomas, M. F. (2000). Late Quaternary environmental changes and the alluvial record in humid tropical environments. *Quaternary International*, **72**, 23–36.

Thorbjarnarson, J., Platt, S. G. & Khaing, U. S. T. (2000). A population survey of the estuarine crocodile in the Ayeyarwady Delta, Myanmar. *Oryx*, **34**, 317–24.

Tompkins, D. M. (1999). Impact of nest-harvesting on the reproductive success of black-nest swiftlets *Aerodramus maximus*. *Wildlife Biology*, **5**, 33–6.

Turner, I. M. (2001). *The Ecology of Trees in the Tropical Rain Forest*. Cambridge: Cambridge University Press.

Turner, I. M., Tan, H. T. W., Wee, Y. C. *et al.* (1994). A study of plant species extinction in Singapore: lessons of the conservation of tropical biodiversity. *Conservation Biology*, **8**, 705–12.

Turner, I. M., Chua, K. S., Ong, J. S. Y., Soong, B. C. & Tan, H. T. W. (1996). A century of plant species loss from an isolated fragment of lowland tropical rain forest. *Conservation Biology*, **10**, 1229–44.

Turner, I. M., Wong, Y. K., Chew, P. T. & Ibrahim, A. B. (1997). Tree species richness in primary and old secondary tropical forest in Singapore. *Biodiversity and Conservation*, **6**, 537–43.

Uhl, C. (1998). Perspectives on wildfire in the humid tropics. *Conservation Biology*, **12**, 942–3.

Valli, E. & Summers, D. (1990). Nest gatherers of Tiger Cave. *National Geographic*, **177**, 107–33.

van Balen, S. (1999). *Differential extinction patterns in javan forest birds*. Birds on fragmented islands. Ph. D. thesis. *Also published as Tropical Resource Management Papers No. 30*. The Netherlands: Wageningen University and Research Centre.

van Balen, S., Nijman, V. & Sozer, R. (1999). Distribution and conservation of the Javan Hawk-eagle *Spizaetus bartelsi*. *Bird Conservation International*, **9**, 333–49.

van Balen, S. B., Dirgayusa, I. W. A., Putra, I. M. W. A. & Prins, H. H. T. (2000). Status and distribution of the endemic Bali starling *Leucopsar rothschildi*. *Oryx*, **34**, 188–97.

van den Bergh, G. D., de Vos, J. & Sondaar, P. Y. (2001). The Late Quaternary palaeogeography of mammal evolution in the Indonesian Archipelago. *Palaeogeography,*

Palaeoclimatology, Palaeoecology, **171**, 385–408.

van derr Kaay, H. J. (1998). Human diseases in relation to the degradation of tropical rainforests. Rainforest Medical Bulletin, **5**, (2).

van Nieuwstadt, M. G. L., Sheil, D. & Kartawinata, K. (2001). The ecological consequences of logging in the burned forests of east Kalimantan, Indonesia. Conservation Biology, **15**, 1183–6.

van Schaik, C. P. (2002). Fragility of traditions: the disturbance hypothesis for the loss of local traditions in orangutans. International Journal of Primatology, **23**, 527–38.

van Schaik, C. P., Monk, K. A. & Robertson, J. M. Y. (2001). Dramatic decline in orangutan numbers in the Leuser Ecosystem, northern Sumatra. Oryx, **35**, 14–25.

Vitousek, P. M., D'Antonio, C. M., Loope, L. L. & Westbrooks, R. (1996). Biological invasions as global environmental change. American Scientist, **84**, 468–78.

von Hippel, F. A. & von Hippel, W. (2002). Sex, drugs and animal parts: will viagra save threatened species? Environmental Conservation, **29**, 277–81.

Voris, H. K. (2000). Maps of Pleistocene sea levels in Southeast Asia: shorelines, river systems and time durations. Journal of Biogeography, **27**, 1153–67.

Voris, H. K. & Inger, R. F. (1995). Frog abundance along streams in Bornean forests. Conservation Biology, **9**, 679–83.

Wallace, A. R. (1876). The Geographical Distribution of Animals with a Study of Relations of Living and Extinct Faunas as Elucidating the Past Changes of the Earth's Surface with Maps and Illustrations. London: MacMillan.

Waltert, M., Mardiastuti, A. & Muhlenberg, M. (2004). Effect of land use on bird species richness in Sulawesi, Indonesia. Conservation Biology, **121**, 1339–46.

Warren, M. S., Hill, J. K., Thomas, J. A. et al. (2001). Rapid responses of British butterflies to opposing forces of climate and habitat change. Nature, **414**, 65–9.

Wells, D. R. (1971). Survival of the Malaysian bird fauna. Malayan Nature Journal, **24**, 248–56.

Wharton, C. H. (1968). Man, fire and wild cattle in Southeast Asia. Proceedings of the Annual Tall Timbers Fire Ecology Conference, **8**, 107–67.

Whitmore, T. C. (1980). The conservation of tropical rainforest. In Conservation Biology:

An Evolutionary–Ecological Perspective, ed. M. E. Soulé & B. A. Wilcox. New York: Sinauer, pp. 308–18.

Whitmore, T. C. (1984). Tropical Rain Forests of the Far East. Oxford: Oxford University Press.

Whitmore, T. C. (1997). Tropical forest disturbance, disappearance, and species loss. In Tropical Forest Remnants: Ecology, Management, and Conservation of Fragmented Communities, ed. W. F. Laurance & R. O. Bierregaard, Jr.. Chicago: University of Chicago Press, pp. 3–12.

Whitmore, T. C. & Sayer, J. A. (1992). Tropical Deforestation and Species Extinction. London: Chapman and Hall.

Whitten, A. J., Mustafa, M. & Henderson, G. S. (1987a). The Ecology of Sulawesi. Yogyakarta, Indonesia: Gadjah Mada University Press.

Whitten, A. J., Damanik, S. J., Anwar, J. & Hisyam, N. (1987b). The Ecology of Sumatra, 2nd edn. Yogyakarta, Indonesia: Gadjah Mada University Press.

Whitten, T. L., Holmes, D. A. & MacKinnon, K. (2001). Conservation biology: a displacement behavior for academia? Conservation Biology, **15**, 1–3.

Wikramanayake, E., Dinerstein, E., Louks, C. et al. (2002). Terrestrial Ecoregions of the Indo-Pacific: a Conservation Assessment. Washington, DC: Island Press.

Willott, S. J. (1999). The effects of selective logging on the distribution of moths in a Bornean rainforest. Philosophical Transactions of the Royal Society of London B Biological Sciences, **354**, 1783–90.

Willott, S. J., Lim, D. C., Compton, S. G. & Sutton, S. L. (2000). Effects of selective logging on the butterflies of a Bornean rainforest. Conservation Biology, **14**, 1055–65.

Wilson, C. C. & Wilson, W. L. (1975). The influence of selective logging on primates and some other animals in east Kalimantan, Indonesia. Folia Primatologica, **23**, 245–74.

Wilson, E. O. (1988). The current state of biological diversity. In Biodiversity, ed. E. O. Wilson & F. M. Peter. Washington, DC: National Academy Press.

Wilson, E. O. (1989). Threats to biodiversity. Scientific American, **261**, 108–16.

Wilson, E. O. (2000). Doomed to early demise. UNESCO Courier (May 2000).

Wilson, M. E. J. & Rosen, B. R. (1998). Implications of paucity of corals in the Paleogene of SE Asia: plate tectonics or centre of origin. In *Biogeography and Geological Evolution of SE Asia*, ed. R. Hall & J. D. Holloway. Leiden: Backhuys, pp. 165–95.

Wilson, W. L. & Johns, A. D. (1982). Diversity and abundance of undisturbed forest, selectively logged forest and plantations in east Kalimantan, Indonesia. *Biological Conservation*, **24**, 205–18.

Wolfe, N. D., Switzer, W. M., Carr, J. K. *et al.* (2004). Naturally acquired simian retrovirus infections in central African hunters. *The Lancet*, **363**, 932–7.

Wong, M. (1985). Understorey birds as indicators of regeneration in a patch of selectively logged west Malaysian rainforest. *International Council for Bird Preservation (ICBP) Technical Publication No. 4*.

Wong, M. (1986). Trophic organization of understorey birds in a Malaysian dipterocarp forest. *Auk*, **103**, 100–16.

Wong, S. T., Servheen, C. W. & Ambu, L. (2004). Home range, movement and activity patterns, and bedding sites of Malayan sun bears *Helarctos malayanus* in the rainforest of Borneo. *Biological Conservation*, **119**, 169–81.

Wong, T. C. M., Sodhi, N. S. & Turner, I. M. (1998). Artificial nest and seed predation experiments in tropical lowland rainforest remnants of Singapore. *Biological Conservation*, **85**, 97–104.

Woodruff, D. S. (2003). Neogene marine transgressions, palaeography and biogeographic transitions on the Thai–Malay Peninsula. *Journal of Biogeography*, **30**, 551–67.

Woods, P. (1989). Effects of logging, drought and fire on structure and composition of tropical forests in Sabah, Malaysia. *Biotropica*, **21**, 290–8.

WorldBank (2003). World Development Indicators 2003. World Bank Group (www.worldbank.org).

World Conservation Monitoring Centre [WCMC] (2002). *Convention on International Trade in Endangered Species of Wild Fauna and Flora (CITES) Annual Report Data*. WCMC CITES Trade Database (www.cites.org).

World Resources Institute [WRI] (2000). *World Resources 2000–2001, People and Ecosystems:*

the Fraying Web of Life. Washington, DC: WRI.

World Resources Institute [WRI] (2003). *World Resources 2002–2004: Decisions for the Earth: Balance, Voice, and Power*. United Nations Development Programme, United Nations Environment Programme, World Bank, World Resources Institute.

Yasuda, M., Ishii, N., Okuda, T. & Hussein, N. A. (2003). Small mammal community: habitat preference and effects after selective logging. In *Pasoh: Ecology of a Lowland Rain Forest of Southeast Asia*, ed. T. Okuda, N. Manokaran, Y. Matsumoto, K. Niyama, S. C. Thomas & P. S. Ashton. Tokyo: Springer-Verlag, pp. 533–46.

Yob, J. M., Field, H., Rashdi, A. M. *et al.* (2001). Nipha virus infection in bats (Order Chiroptera) in Peninsular Malaysia. *Emerging Infectious Diseases*, **7**, 439–41.

Zakaria, M. & Nordin, M. (1998). Comparison of frugivory by birds in primary and logged lowland dipterocarp forests in Sabah, Malaysia. *Tropical Biodiversity*, **5**, 1–9.

Zavaleta, E. S., Shaw, M. R., Chiariello, N. R., Mooney, H. A. & Field, C. B. (2003). Additive effects of simulated climate changes, elevated CO_2, and nitrogen deposition on grassland diversity. *Proceedings of the National Academy of Sciences of the United States of America*, **100**, 7650–4.

Zhang, H., Henderson-Sellers, A. & McGuffie, K. (2001). The compounding effects of tropical deforestation and greenhouse warming on climate. *Climatic Change*, **49**, 309–38.

Zimmerman, P. R., Greenberg, J. P., Wandiga, S. O. & Crutzen, P. J. (1982). Termites: a potentially large source of atmospheric methane, carbon dioxide, and molecular hydrogen. *Science*, **218**, 563–5.

Zubaid, A. & Rizal, M. (1995). The relative abundance of rats between two forest types in Peninsular Malaysia. *Malayan Nature Journal*, **49**, 139–42.

Zuraina, M. (1982). The West Mouth, Niah, in the prehistory of South-east Asia. *Sarawak Museum Journal (special monograph no. 3)*, **31**, 1–20.

Zwiers, F. W. (2002). The 20-year forecast. *Nature*, **416**, 690–1.

INDEX